MW00835248

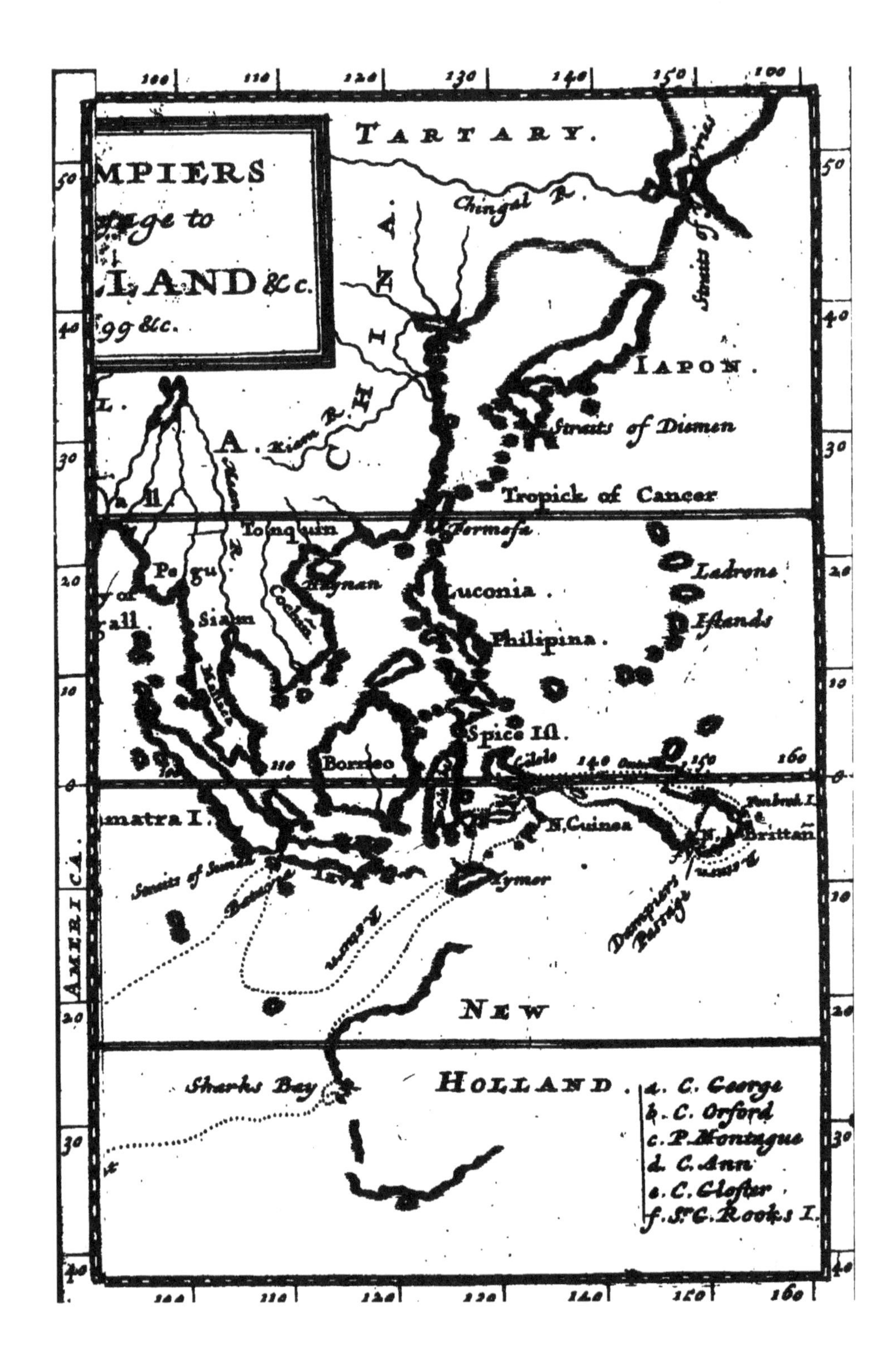

MPIERS
age to
LAND &c.
99 &c.
TARTARY.
Chingal R.
CHINA
IAPON.
Streits of Diemen
Tropick of Cancer
Tonquin
Formosa
Pegu
Siam
Haynan
Cochin
Luconia
Ladrons
Islands
Philipina.
Spice Isl.
Borneo
N. Guinea
N. Brittan
Sumatra I.
Timor
Dampiers Passage
AMERICA.
NEW
HOLLAND.
Sharks Bay
a. C. George
b. C. Orford
c. P. Montague
d. C. Ann
e. C. Gloster
f. Sr. G. Rooks I.

A VOYAGE TO *New Holland*, &c.

In the Year, 1699.

Wherein are deſcribed,

The *Canary*-Iſlands, the Iſles of *Mayo* anc St. *Jago*. The Bay of *All Saints*, with th Forts and Town of *Bahia* in *Braſil*. Cap *Salvadore*. The Winds on the *Braſilia* Coaſt. *Abrohlo*-Shoals. A Table of all th *Variations* obſerv'd in this Voyage. Oc currences near the Cape of *Good Hope* The Courſe to *New Holland*. *Shark*'s Bay The Iſles and Coaſt, *&c.* of *New Holland*.

Their Inhabitants, Manners, Cuſtoms, Trade, *&c* Their Harbours, Soil, Beaſts, Birds, Fiſh, *&c* Trees, Plants, Fruits, *&c.*

Illuſtrated with ſeveral Maps and Draughts; alſc divers Birds, Fiſhes, and Plants, not found ir this part of the World, Curiouſly Ingraven or Copper-Plates.

VOL. III.

By Captain *William Dampier*.

LONDON:

Printed for *James Knapton*, at the *Crown* in St. *Paul*' Church-yard, 1703.

To the Right Honourable

THOMAS

Earl of *Pembroke*,

Lord President of Her Majesty's Most Honourable Privy Council, *&c.*

My Lord,

THE Honour I had of being employ'd in the Service of His late Majesty of Illustrious Memory, at the time when Your Lordship presided at the Admiralty, gives me

 the

the Boldness to ask Your Protection of the following Papers. They consist of some Remarks made upon very distant Climates, which I should have the vanity to think altogether new, cou'd I persuade my self they had escap'd Your Lordship's Knowledge. However I have been so cautious of publishing any thing in my whole Book that is generally known, that I have deny'd my self the pleasure of paying the due Honours to Your Lordships Name in the Dedication. I am asham'd, my Lord, to offer You so imperfect a Present, having not time to set down all the Memoirs of my last Voyage: But as the particular Service I have now undertaken, hinders me from finishing

nishing this Volume, so I hope it will give me an opportunity of paying my Respects to Your Lordship in a new one.

The World is apt to judge of every thing by the Success; and whoever has ill Fortune will hardly be allow'd a good Name. This, my Lord, was my Unhappiness in my late Expedition in the Roe-Buck, *which founder'd thro' perfect Age near the Island of* Ascension. *I suffer'd extreamly in my Reputation by that Misfortune; tho' I comfort my self with the Thoughts, that my Enemies cou'd not charge any Neglect upon me. And since I have the Honour to be acquitted by Your Lordship's Judgment, I shou'd be very humble not to value my ſelf upon ſo compleat a Vindication. This, and a World of other Favours, which I have been ſo happy as to receive from Your Lordſhip's Goodneſs, do engage me to be with an everlaſting Reſpect,*

My Lord,

Your Lordſhip's

Moſt Faithful and

Obedient Servant,

Will. Dampier.

THE PREFACE.

THE favourable Reception my two former Volumes of *Voyages and Descriptions* have already met with in the World, gives me Reason to hope, That notwithstanding the Objections which have been raised against me by prejudiced Persons, this *Third Volume* likewise may in some measure be acceptable to Candid and Impartial Readers, who are curious to know the Nature of the Inhabitants, Animals, Plants, Soil, *&c.* in those distant Countries, which have either seldom or not at all been visited by any *Europeans.*

It has almoſt always been the Fate of thoſe who have made new Diſcoveries, to be diſeſteemed and ſlightly ſpoken of, by ſuch as either have had no true Reliſh and Value for the *Things themſelves* that are diſcovered, or have had ſome Prejudice againſt *the Perſons* by whom the Diſcoveries were made. It would be vain therefore and unreaſonable in me to expect to eſcape the Cenſure of all, or to hope for better Treatment than far Worthier Perſons have met with before me. But this Satisfaction I am ſure of having, that the *Things themſelves* in the Diſcovery of which I have been imployed, are moſt worthy of our Diligenteſt Search and Inquiry; being the various and wonderful Works of God in different Parts of the World: And however *unfit a Perſon* I may be in other reſpects to have undertaken this Task, yet at leaſt I have given a faithful Account,

count, and have found *some* Things undiſcovered by any before, and which may at leaſt be *some* Aſſiſtance and Direction to better qualified Perſons who ſhall come after me.

It has been Objected againſt me by ſome, that my Accounts and Deſcriptions of Things are dry and jejune, not filled with variety of pleaſant Matter, to divert and gratify the Curious Reader. How far this is true, I muſt leave to the World to judge. But if I have been exactly and ſtrictly careful to give only *True* Relations and Deſcriptions of Things (as I am ſure I have;) and if my Deſcriptions be ſuch as may be of uſe not only to my ſelf (which I have already in good meaſure experienced) but alſo to others in future Voyages; and likewiſe to ſuch Readers at home as are more deſirous of a Plain and Juſt Account of the true Nature and State of the Things deſcribed, than of a Polite and Rhetorical Narrative: I hope all the Defects in my Stile, will meet with an eaſy and ready Pardon.

Others have taxed me with borrowing from other Men's Journals; and with Inſufficiency, as if I was not my ſelf the Author of what I write, but publiſhed Things digeſted and drawn up by others. As to the firſt Part of this Objection, I aſſure the Reader, I have taken nothing from any Man without mentioning his Name, except ſome very few Relations and particular Obſervations received from credible Perſons who deſired not to be named; and theſe I have always expreſly diſtinguiſhed in my Books, from what I relate as of my own obſerving. And as to the latter; I think it ſo far from being a Diminution to one of my Education and Employment, to have what I write, Reviſed and Corrected by Friends; that on the con-

contrary, the beſt and moſt eminent Authors are not aſhamed to own the ſame Thing, and look upon it as an Advantage.

Laſtly, I know there are ſome who are apt to ſlight my Accounts and Deſcriptions of Things, as if it was an eaſie Matter and of little or no Difficulty to do all that I have done, to viſit little more than the Coaſts of unknown Countries, and make ſhort and imperfect Obſervations of Things only near the Shore. But whoever is experienced in theſe Matters, or conſiders Things impartially, will be of a very different Opinion. And any one who is ſenſible, how backward and refractory the Seamen are apt to be in long Voyages when they know not whither they are going, how ignorant they are of the Nature of the Winds and the ſhifting Seaſons of the Monſoons, and how little even the Officers themſelves generally are

skilled

skilled in the Variation of the Needle and the Uſe of the Azimuth Compaſs; beſides the Hazard of all outward Accidents in ſtrange and unknown Seas: Any one, I ſay, who is ſenſible of theſe Difficulties, will be much more pleaſed at the Diſcoveries and Obſervations I have been able to make, than diſpleaſed with me that I did not make more.

Thus much I thought neceſſary to premiſe in my own Vindication, againſt the Objections that have been made to my former Performances. But not to trouble the Reader any further with Matters of this Nature; what I have more to Offer, ſhall be only in relation to the following Voyage.

For the better apprehending the Courſe of this Voyage, and the Situation of the Places mentioned in it, I have here, as in the former Volumes, cauſed a Map to be Ingraven, with a prick'd Line, re-

preſenting

presenting to the Eye the whole Thread of the Voyage at one View; besides Draughts and Figures of particular Places, to make the Descriptions I have given of them more intelligible and useful.

Moreover, which I had not the opportunity of doing in my former Voyages; having now had in the Ship with me a Person skill'd in Drawing, I have by this means been enabled, for the greater Satisfaction of the Curious Reader, to present him with exact Cuts and Figures of several of the principal and most remarkable of those Birds, Beasts, Fishes and Plants, which are described in the following Narrative; and also of several, which not being able to give any better or so good an Account of, as by causing them to be exactly Ingraven, the Reader will not find any further Description of them, but only that they were

found

found in ſuch or ſuch particular Countries. The Plants themſelves are in the Hands of the Ingenious Dr. *Woodward*. I could have cauſed many others to be drawn in like manner, but that I reſolved to confine my Self to ſuch only, as had ſome very remarkable difference in the ſhape of their principal Parts from any that are found in *Europe*. I have beſides ſeveral Birds and Fiſhes ready drawn, which I could not put into the preſent Volume, becauſe they were found in Countries, to the Deſcription whereof the following Narrative does not reach. For, being obliged to prepare for another Voyage, ſooner than I at firſt expected; I have not been able to Continue the enſuing Narrative any further than to my Departure from the Coaſt of *New Holland*. But, if it pleaſe God that I return again ſafe, the Reader may expect a Continuation of this Voyage from my departure from *New Holland*, till the foundring of my Ship near the Island of *Ascension*.

In the mean time, to make the Narrative in some measure compleat, I shall here add a Summary Abstract of that latter part of the Voyage, whereof I have not had time to draw out of my Journals a full and particular Account at large. Departing therefore from the Coast of *New Holland* in the beginning of *September*, 1699. (for the Reasons mentioned *Page* 154.) we arrived at *Tymor*, *Sept*. 15. and Anchored off that Island. On the 24th we obtain'd a small Supply of fresh Water from the Governor of a *Dutch* Fort and Factory there; we found also there a *Portuguese* Settlement, and were kindly treated by them. On the 3d of *December* we arrived on the Coast of *New Guinea*; where we found good fresh Water, and had Commerce with the Inhabitants of a certain

certain Island call'd *Pulo-Sabuti*. After which, passing to the Northward, we ranged along the Coast to the Eastermost part of *New Guinea*: which I found does not join to the main Land of *New-Guinea*, but is an Island, as I have described it in my Map; and call'd it *New Britain*.

It is probable this Island may afford many rich Commodities, and the Natives may be easily brought to Commerce. But the many Difficulties I at this time met with, the want of convenience to clean my Ship, the fewness of my Men, their desire to hasten home, and the danger of continuing in these Circumstances in Seas where the Shoals and Coasts were utterly unknown, and must be searched out with much Caution and length of time; hindred me from prosecuting any further at present my intended Search. What I have been able to do in this Matter

ter for the Publick Service, will, I hope, be candidly receiv'd; and no Difficulties ſhall diſcourage me from endeavouring to promote the ſame End, whenever I have an opportunity put into my Hands.

May 18. in our return, we arrived again at *Tymor*. *June* 21, we paſt by part of the Iſland *Java*. *July* 4, we anchored in *Batavia*-Road; and I went aſhore, viſited the *Dutch* General, and deſired the Privilege of buying Proviſions that I wanted; which was granted me. In this Road we lay till the 17th of *October* following; when, having fitted the Ship, recruited my Self with Proviſions, filled all my Water, and the Seaſon of the year for returning towards *Europe* being come; I ſet Sail from *Batavia*, and on the 19th of *December* made the Cape of *Good Hope*; whence departing *Jan.* 11, we made the Iſland of *Santa Hellena* on the 31ſt; and *February* the 21ſt. the Iſland

 of

of *Ascension*; near to which my Ship, having sprung a Leak which could not be stopped, foundred at Sea; With much difficulty we got ashore, where we liv'd on Goats and Turtle; and on the 26th of *February* found, to our great Comfort, on the S. E. side of a high Mountain, about half a mile from its top, a Spring of fresh Water. I returned to *England* in the *Canterbury East-India*-Ship. For which wonderful Deliverance from so many and great Dangers, I think my self bound to return continual Thanks to Almighty God; whose Divine Providence if it shall please to bring me safe again to my Native Country from my present intended Voyage; I hope to publish a particular Account of all the material Things I observed in the several Places which I have now but barely mentioned.

THE CONTENTS.

CHAP. I.

 Salt,

CHAP. II.

CHAP.

CHAP. III.

Dampier's Voyages.

VOL. III.

A Voyage to Terra Auſtralis.

CHAP. I.

The A.'s departure from the Downs. *A Caution to thoſe who Sail in the Channel. His Arrival at the* Canary-Iſlands. Santa Cruz *in* Teneriffe; *the Road and Town, and* Spaniſh *Wreck.* Laguna *T. Lake and Country; and* Oratavia *T. and Road. Of the Wines and other Commodities of* Teneriffe, &c. *and the Governors at* Laguna *and* Santa Cruz. *Of the Winds in theſe Seas. The A.'s Arrival*

An. 1699.

rival at Mayo, *one of the C.* Verd Islands; *its Salt-pond, compar'd with that of* Salt-Tortuga; *its Trade for Salt, and* Frape-*boats. Its Vegetables, Silk-Cotton,* &c. *Its Soil, and Towns; its* Guinea-*Hen's, and other Fowls, Beasts, and Fish. Of the Sea-Turtle's* (&c.) *laying in the Wet Season. Of the Natives, their Trade and Livelihood. The A.'s Arrival at J.* St. Jago; *and* St. Jago *Town. Of the Inhabitants, and their Commodities. Of the Custard-Apple, and the* Papah. *St.* Jago *Road.* J. Fogo.

I Sail'd from the *Downs* early on *Saturday* *Jan.* 14. 169$\frac{8}{9}$. with a fair Wind, in his Majesty's Ship the *Roe-buck*; carrying but 12 Guns in this Voyage, and 50 Men and Boys, with 20 Month's Provision. We had several of the King's Ships in Company, bound for *Spit-head* and *Plimouth*; and by Noon we were off *Dungeness*. We parted from them that Night, and stood down the Channel, but found our Selves next Morning nearer the *French* Coast than we expected; C. *de Hague* bearing S. E. and by E. 6 L. There were many other Ships, some nearer, some further off the *French* Coast, who all seem'd seem'd to have gone nearer to it than they thought they should. My Master, who was somewhat troubled at it at first, was not displeas'd however to find that he had Company in his Mistake: Which, as I have heard, is a very common one, and fatal to many Ships. The Occasion of it is the not allowing for the Change of the Variation since the making of the Charts; which Captain *Halley* has observ'd to be very considerable. I shall refer the Reader to his own Account of it which he caus'd to be Publish'd in a single Sheet of Paper, purposely for a Caution to such as pass to and fro the *English* Channel: The Title of it is in the Margin. And my own Experience thus confirming to me the Usefulness of such a Caution, I was willing to take this occasion of helping towards the making it the more Publick.

An Advertisement necessary to be observ'd in the Navigation up and down the Channel of England. Sold by *S. Smith* at the *Prince's Arms* in St. *Paul's* Church-yard. Price 2 *d.*

Not to trouble the Reader with every Days Run, nor with the Winds or Weather (but only in the remoter Parts, where it may be more particularly useful) standing away from C. *la Hague*, we made the *Start* about 5 that Afternoon; which being the last Land we saw of *England*, we reckon'd our Departure from thence: Tho' we had rather have taken it from the *Lizard*, if the hazy Weather would have suffer'd us to have seen it.

The first Land we saw after we were out of the Channel was C. *Finisterre*, which we made on the 19th; and on the 28th made

An. 1699. *Lancerota*, one of the *Canary* Islands; of which, and of *Allegrance*, another of them; I have here given the *Sights*, as they both appeared to us at two several Bearings and Distances. [Table I. N°. 1, 2.]

We were now standing away for the Island *Teneriffe*, where I intended to take in some Wine and Brandy for my Voyage. On *Sunday*, half an hour past 3 in the Afternoon, we made the Island, and crouded in with all our Sails till 5; when the N. E. Point of the Isle bore W. S. W. dist. 7 Leagues: But being then so far off that I could not expect to get in before Night, I lay by till next Morning, deliberating whether I should put in at *Santa Cruz*, or at *Oratavia*, the one on the E. the other on the W. side of the Island; which lies mostly North and South; and these are the principal Ports on each Side. I chose *Santa Cruz* as the better Harbour (especially at this time of the Year) and as best furnish'd with that sort of Wine which I had occasion to take in for my Voyage: So there I come to an Anchor *Jan.* 30th, in 33 Fathom-water, black slimy Ground; about half a Mile from the Shore; from which distance I took the Sight of the Town. [Table I. N°. 3.]

In this Road Ships must ride in 30, 40, or 50 Fathom-water, not above half a mile from the Shore at farthest: And if there are many Ships, they must ride close one by another. The Shore is generally high Land, and in most Places steep to. This Road lies so open

to

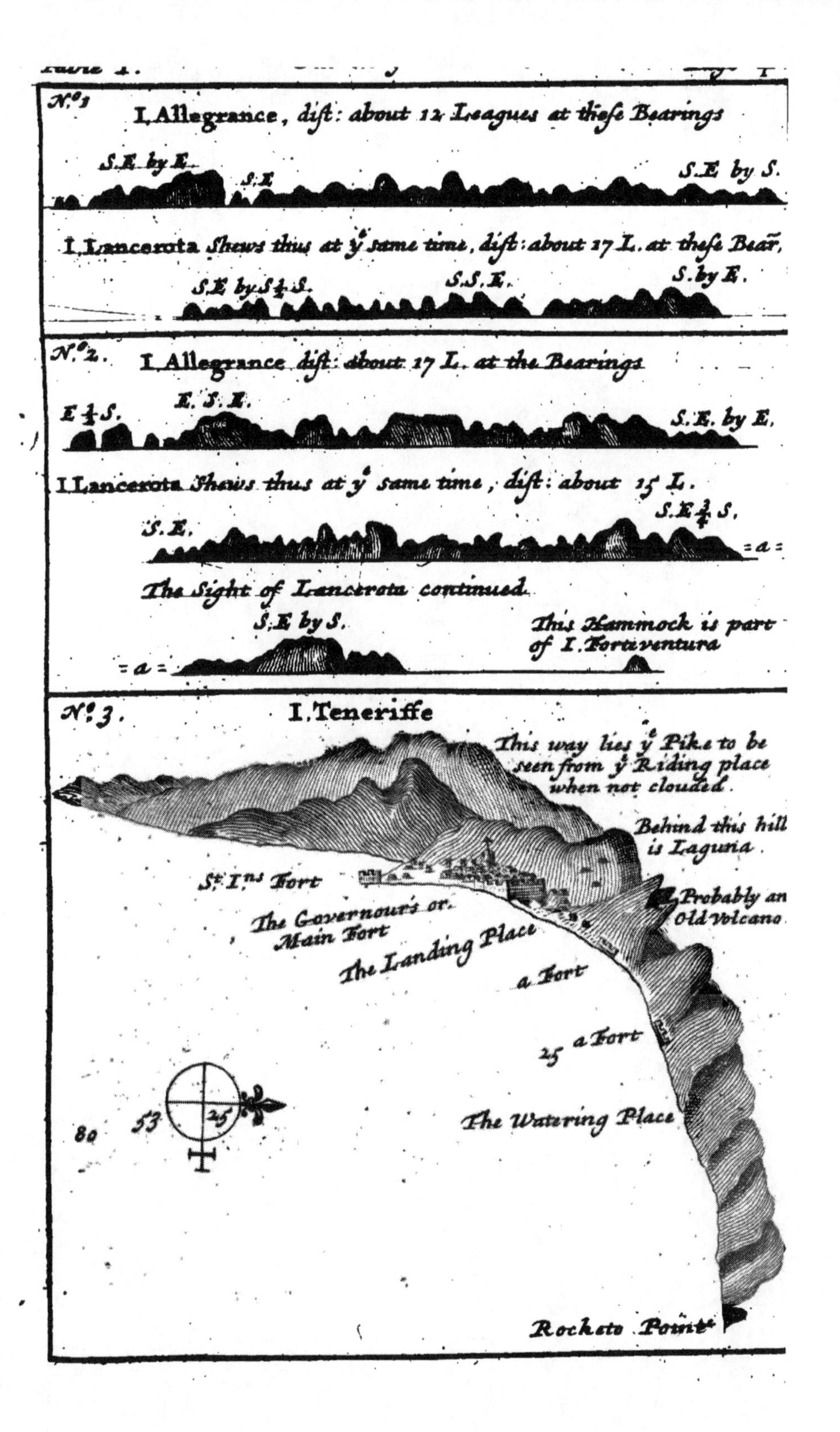
N.° 1
I. Allegrance, dist: about 12 Leagues at these Bearings
S.E. by E.
S.E
S.E by S.
I. Lancerota Shews thus at ye same time, dist: about 17 L. at these Bear.
S.E by S ½ S.
S.S.E.
S. by E.
N.° 2.
I. Allegrance dist: about 17 L. at the Bearings
E ½ S.
E. S. E.
S. E. by E.
I. Lancerota Shews thus at ye same time, dist: about 15 L.
S. E.
S.E ¾ S.
= a =
The Sight of Lancerota continued
S.E by S.
= a =
This Hammock is part of I. Forteventura
N.° 3.
I. Teneriffe
This way lies ye Pike to be seen from ye Riding place when not clouded.
Behind this hill is Laguna.
St. I.ns Fort
Probably an Old Volcano
The Governour's or Main Fort
The Landing Place
a Fort
25
a Fort
80
53
25
The Watering Place
Rocketo Point

to the East, that Winds from that side make a great Swell, and very bad going ashore in Boats: The Ships that ride here are then often forced to put to Sea, and sometimes to cut or slip their Anchors, not being able to weigh them. The best and smoothest Landing is in a small sandy Cove, about a mile to the N. E. of the Road, where there is good Water, with which Ships that lade here are supply'd; and many times Ships that lade at *Oratavia*, which is the chief Port for Trade, send their Boats hither for Water. That is a worse Port for Westerly than this is for Easterly Winds; and then all Ships that are there put to Sea. Between this Watering-place and *Santa Cruz* are two little Forts; which with some Batteries scatter'd along the Coast command the Road. *Santa Cruz* its self is a small unwalled Town fronting the Sea, guarded with two other Forts to secure the Road. There are about 200 Houses in the Town, all 2 Stories high, strongly built with Stone, and covered with Pantile. It hath two Convents and one Church, which are the best Buildings in the Town. The Forts here could not secure the *Spanish* Galleons from Admiral *Blake*, tho' they hall'd in close under the main Fort. Many of the Inhabitants that are now living remember that Action; in which the *English* batter'd the Town, and did it much Damage; and the marks of the Shot still remain in the Fort-Walls. The Wrecks of the Galleons that

 were

*An.*1699. were burnt here lie in 15 Fathom-water: And 'tis said that most of the Plate lies there, tho' some of it was hastily carried ashore at *Blake*'s coming in sight.

Soon after I had anchor'd I went ashore here to the Governor of the Town, who received me very kindly and invited me to Dine with him the next day. I return'd on board in the Evening, and went ashore again with two of my Officers the next Morning; hoping to get up the Hill time enough to see *Laguna*, the principal Town, and to be back again to Dine with the Governor of *Santa Cruz*; for I was told that *Laguna* was but 3 Mile off. The Road is all the way up a pretty steep Hill; yet not so steep but that Carts go up and down laden. There are Publick Houses scattering by the way-side, where we got some Wine. The Land on each side seemed to be but rocky and dry; yet in many Places we saw Spots of green flourishing Corn. At farther distances there were small Vineyards by the Sides of the Mountains, intermixt with abundance of waste rocky Land, unfit for Cultivation, which afforded only Dildo-bushes. It was about 7 or 8 in the Morning when we set out from *Santa Cruz*; and it being fair clear Weather, the Sun shone very bright and warmed us sufficiently before we got to the City *Laguna*; which we reached about 10 a Clock, all sweaty and tired, and were glad to refresh our selves with a little Wine in a sorry Tipling-house:

But

But we soon found out one of the *English* Merchants that resided here; who entertain'd us handsomly at Dinner, and in the Afternoon shew'd us the Town. *An.* 1699.

Laguna is a pretty large well-compacted Town, and makes a very agreeable Prospect. It stands part of it against a Hill, and part in a Level. The Houses have mostly strong Walls built with Stone and covered with Pantile. They are not uniform, yet they appear pleasant enough. There are many fair Buildings; among which are 2 Parish-Churches, 2 Nunneries, an Hospital, 4 Convents, and some Chapels; besides many Gentlemen's Houses. The Convents are those of St. *Austin*, St. *Dominick*, St. *Francis*, and St. *Diego*. The two Churches have pretty high square Steeples, which top the rest of the Buildings. The Streets are not Regular, yet they are mostly spacious and pretty handsome; and near the middle of the Town is a large Parade, which has good Buildings about it. There is a strong Prison on one side of it; near which is a large Conduit of good Water, that supplies all the Town. They have many Gardens which are set round with Oranges, Limes, and other Fruits: In the middle of which are Pot-herbs, Sallading, Flowers, *&c.* And, indeed, if the Inhabitants were curious this way, they might have very pleasant Gardens: For as the Town stands high from the Sea, on the Brow of a Plain that is all open to the East, and

An. 1699. hath consequently the Benefit of the true Trade-wind, which blows here, and is most commonly fair; so there are seldom wanting, at this Town, brisk, cooling, and refreshing Breezes all the Day.

On the back of the Town there is a large Plain of 3 or 4 Leagues in length and 2 Miles wide, producing a thick kindly sort of Grass, which look'd green and very pleasant when I was there, like our Meadows in *England* in the Spring. On the East-side of this Plain, very near the back of the Town, there is a natural Lake or Pond of fresh Water. It is about half a Mile in circumference; but being stagnant, 'tis only us'd for Cattle to drink of. In the Winter-time several sorts of wild Fowl resort hither, affording plenty of Game to the Inhabitants of *Laguna*. This City is called *Laguna* from hence; for that Word in *Spanish* signifies a Lake or Pond. The Plain is bounded on the W. the N. W. and the S. W. with high steep Hills; as high above this Plain as this is above the Sea; and 'tis from the foot of one of these Mountains that the Water of the Conduit which supplies the Town, is conveyed over the Plain, in Troughs of Stone rais'd upon Pillars. And, indeed, considering the Situation of the Town, its large Prospect to the East (for from hence you see the *Grand Canary)* its Gardens, cool Arbors, pleasant Plain, green Fields, the Pond and Aqueduct, and its refreshing Breezes, it is a very delightful Dwelling; espe-

especially for such as have not Business that calls them far and often from home: For the Island being generally Mountainous, steep and craggy, full of Risings and Fallings, 'tis very troublesome Travelling up and down in it, unless in the Cool of the Mornings and Evenings; And Mules and Asses are most us'd by them, both for Riding and Carriage, as fittest for the stony, uneven Roads.

An. 1699.

Beyond the Mountains, on the S. W. side, still further up, you may see from the Town and Plain a small peeked Hill, overlooking the rest. This is that which is called the *Pike of Teneriffe*, so much noted for its heighth: But we saw it here at so great a disadvantage, by reason of the nearness of the adjacent Mountains to us, that it looked inconsiderable in respect to its Fame.

The true *Malmesy* Wine grows in this Island; and this here is said to be the best of its kind in the World. Here is also *Canary*-Wine, and *Verdona*, or Green-wine. The *Canary* grows chiefly on the West-side of the Island; and therefore is commonly sent to *Oratavia*; which being the chief Sea-port for Trade in the Island, the principal *English* Merchants reside there, with their Consul; because we have a great Trade for this Wine. I was told, That that Town is bigger than *Laguna*; that it has but one Church, but many Convents: That the Port is but ordinary at best, and is very bad when the N. W. Winds blow. These Norwesters give notice

*An.*1699. tice of their coming, by a great Sea that tumbles in on the Shore for ſome time before they come, and by a black Sky in the N. W. Upon theſe Signs Ships either get up their Anchors, or ſlip their Cables and put to Sea, and ply off and on till the Weather is over. Sometimes they are forced to do ſo 2 or 3 times before they can take in their Lading; which 'tis hard to do here in the faireſt Weather: And for freſh Water, they ſend, as I have ſaid, to *Santa Cruz*. *Verdona* is green, ſtrong-bodied Wine, harſher and ſharper than *Canary*. 'Tis not ſo much eſteemed in *Europe*, but is exported to the *Weſt-Indies*, and will keep beſt in hot Countries; for which Reaſon I touch'd here to take in ſome of it for my Voyage. This ſort of Wine is made chiefly on the Eaſt-ſide of the Iſland, and Shipt off at *Santa Cruz*.

Beſides theſe Wines, which are yearly vended in great plenty from the *Canary* Iſlands. (chiefly from *Grand Canary*, *Teneriffe*, and *Palma*) here is ſtore of Grain, as Wheat, Barly and Maiz, which they often tranſport to other places. They have alſo ſome Beans and Peas, and Coches, a ſort of Grain much like Maiz, ſow'd moſtly to fatten Land. They have Papah's, which I ſhall ſpeak more of hereafter; Apples, Pears, Plumbs, Cherries, and excellent Peaches, Apricocks, Guava's, Pomegranates, Citrons, Oranges, Lemons, Limes, Pumpkins, Onions the beſt in the World, Cabbages, Turnips, Potato's, *&c*.
They

They are also well stocked with Horses, Cows, Asses, Mules, Sheep, Goats, Hogs, Conies, and plenty of Deer. The *Lancerot* Horses are said to be the most mettlesome, fleet, and loyal Horses that are. Lastly, here are many Fowls, as Cocks and Hens, Ducks, Pidgeons, Patridges, *&c.* with plenty of Fish, as Mackril, *&c.* All the *Canary* Islands have of these Commodities and Provisions more or less: But as *Lancerota* is most fam'd for Horses; and *Grand Canary*, *Teneriffe*, and *Palma* for Wines, *Teneriffe* especially for the best Malmesy, (for which reason these 3 Islands have the chief Trade) so is *Forteventura* for Dunghil-Fowls, and *Gomera* for Deer. Fowls and other Eatables are dear on the Trading Islands; but very plentiful and cheap on the other; and therefore 'tis best for such Ships as are going out on long Voyages, and who design to take in but little Wine, to touch rather at these last; where also they may be supply'd with Wine enough, and good cheap: And for my own part, if I had known it before I came hither, I should have gone rather to one of those Islands than to *Teneriffe*: But enough of this.

An. 1699

'Tis reported they can raise 12000 armed Men on this Island. The Governor or *General* (as he is call'd) of all the *Canary* Islands lives at *Laguna*: His Name is *Don Pedro de Ponto*. He is a Native of this Island, and was not long since President of *Panama* in the *South Seas*; who bringing some very rich

Pearls

An. 1699. Pearls from thence, which he presented to the Queen of *Spain*, was therefore, as 'tis said, made General of the *Canary* Islands. The *Grand Canary* is an Island much superior to *Teneriffe* both in Bulk and Value; but this Gentleman chuses rather to reside in this his native Island. He has the Character of a very worthy Person; and governs with Moderation and Justice, being very well beloved.

One of his Deputies was the Governor of *Santa Cruz*, with whom I was to have Din'd; but staying so long at *Laguna*, I came but time enough to Sup with him. He is a civil, discreet Man. He resides in the main Fort close by the Sea. There is a Centinel stands at his Door; and he has a few Servants to wait on him. I was Treated in a large dark Lower Room, which has but one small Window. There were about 200 Muskets hung up against the Walls, and some Pikes; no Wainscot, Hangings, nor much Furniture. There was only a small old Table, a few old Chairs, and 2 or 3 pretty long Forms to sit on. Having Supp'd with him, I invited him on Board, and went off in my Boat. The next Morning he came aboard with another Gentleman in his Company, attended by 2 Servants: But he was presently Sea-sick, and so much out of order that he could scarce Eat or Drink any Thing, but went quickly ashore again.

Having

Having refresh'd my Men ashore, and taken in what we had occasion for, I Sail'd away from *Santa Cruz* on *Feb.* 4. in the Afternoon; hastening out all I could, because the N. E. Winds growing stormy made so great Sea, that the Ship was scarce safe in the Road; and I was glad to get out, tho' we left behind several Goods we had bought and paid for: For a Boat could not go ashore; and the stress was so great in weighing Anchor, that the Cable broke. I design'd next for the I. of *Mayo*, one of the C. *Verd* Islands; and ran away with a strong N. E. Wind, right afore it, all that Night and the next Day, at the rate of 10 or 11 Miles an hour; when it slacken'd to a more moderate Gale. The *Canary* Islands are, for their Latitude, within the usual Verge of the True or General Trade-Wind; which I have observ'd to be, on this side the Equator, N. Easterly: But then lying not far from the *African* Shore, they are most subject to a N. Wind, which is the *Coasting and constant Trade*, sweeping that Coast down as low as to C. *Verd*; which spreading in breadth, takes in mostly the *Canary* Islands; tho' it be there interrupted frequently with the True Trade-Wind, N. West-Winds, or other Shifts of Wind that Islands are Subject to; especially where they lie many together. The *Pike* of *Teneriffe*, which had generally been Clouded while we lay at *Santa Cruz*, appear'd now all white with Snow, hovering over the other Hills; but

An. 1699.

An. 1699. but their heighth made it seem the less considerable; for it looks most remarkable to Ships that are to the Westward of it. We had brisk N. N. E. and N. E. Winds from Teneriffe; and saw Flying-fish, and a great deal of Sea-thistle Weed floating. By the 9th of Feb. at Noon we were in the Lat. of 15 d. 4 m. so we steered away W. N. W. for the I. of Mayo, being by Judgment, not far to the E. of it, and at 8 a Clock in the Evening lay by till Day. The Wind was then at W. by S. and so it continued all Night, fair Weather, and a small easie Gale. All these were great Signs, that we were near some Land, after having had such constant brisk Winds before. In the Morning after Sun-rise, we saw the Island at about 4 Leagues distance: But it was so hazy over it, that we could see but a small part of it; yet even by that part I knew it to be the Isle of Mayo. See how it appear'd to us at several Views, as we were compassing the E. the S. E. and the S. of it, to get to the Road, on the S. W. of it, [Table II. N°. 1, 2, 3.] and the Road it self [N°. 4.]

I got not in till the next Day, Feb. 11. when I come to an Anchor in the Road, which is the Lee-ward part of the Island; for 'tis a general Rule never to Anchor to Wind-ward of an Island between the Tropicks. We Anchored at 11 a Clock in 14 Fathom clean Sand, and very smooth Water, about three quarters of a Mile from the Shore,

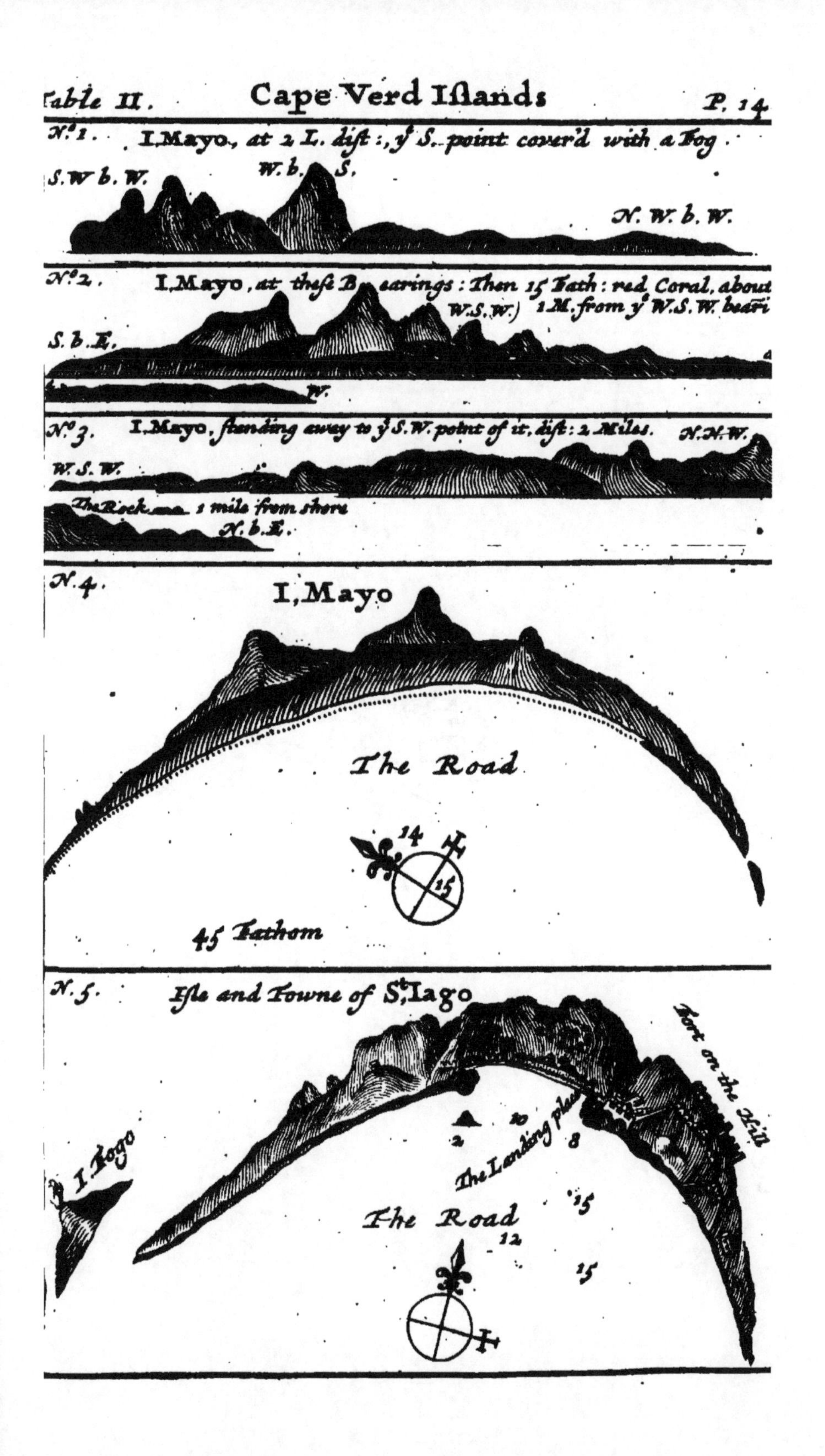
Table II.
Cape Verd Islands
P. 14
N.º 1. I. Mayo, at 2 L. dist:, ye S. point cover'd with a Fog
S.W b. W.
W. b. S.
N. W. b. W.
N.º 2. I. Mayo, at these Bearings: Then 15 Fath: red Coral, about
W.S.W.) 1 M. from ye W.S.W. beari
S. b. E.
W.
N.º 3. I. Mayo, standing away to ye S.W. point of it, dist: 2 Miles.
N.N.W.
W. S. W.
The Rock 1 mile from shore
N. b. E.
N. 4.
I, Mayo
The Road
14
15
45 Fathom
N. 5.
Isle and Towne of St. Iago
Fort on the Hill
The Landing place
10
2
8
I. Fogo
15
The Road
12
15

Shore, in the same Place where I Anchor'd in my *Voyage round the World*; and found riding here the *Newport of London*, a Merchant Man, Captain *Barefoot* Commander, who welcomed me with 3 Guns, and I returned one for Thanks. He came from *Fayal*, one of the *Western* Islands; and had store of Wine and Brandy aboard. He was taking in Salt to carry to *New-found-Land*, and was very glad to see one of the King's Ships, being before our coming afraid of Pyrates; which, of late Years, had much infested this and the rest of the Cape *Verd Islands*.

An. 1699.

I have given some Account of the Island of *Mayo*, and of other of these Islands, in my *Voyage round the World*, [Vol. I. p. 70.] but I shall now add some further Observations that occurr'd to me in this Voyage. The I. of *Mayo* is about 7 Leagues in Circumference, of a roundish Form, with many small rocky Points shooting out into the Sea a Mile, or more. Its Lat. is 15 d. N. and as you Sail about the Isle, when you come pretty nigh the Shore, you will see the Water breaking off from those Points; which you must give a Birth to, and avoid them. I Sail'd at this time two parts in three round the Island, but saw nothing dangerous besides these Points; and they all shew'd themselves by the Breaking of the Water: Yet 'tis reported, That on the N. and N. N. W. side there are dangerous Sholes, that ly farther off at Sea; but I was not on that Side. There are 2 Hills

on

An. 1699. on this Island of a considerable heighth; one pretty bluff, the other peeked at top. The rest of the Island is pretty level, and of a good heighth from the Sea. The Shore clear round hath sandy Bays, between the Rocky Points I spake of; and the whole Island is a very dry sort of Soil.

On the West-side of the Isle where the Road for Ships is there is a large Sandy Bay, and a Sand-bank, of about 40 Paces wide within it, which runs along the Shore 2 or 3 Miles; within which there is a large *Salina* or Salt-pond, contained between the Sand-bank and the Hills beyond it. The whole *Salina* is about 2 Miles in length, and half a Mile wide; but above one half of it is commonly dry. The North end only of the Pond never wants Water, producing Salt from *November* till *May*, which is here the dry Season of the Year. The Water which yields this Salt works in from out of the Sea through a hole in the Sand-bank before mentioned, like a Sluce, and that only in Spring-tides; when it fills the Pond more or less, according to the heighth of the Tides. If there is any Salt in the Ponds when the Flush of Water comes in, it presently dissolves: But then in two or three Days after it begins to Kern; and so continues Kerning till either all, or the greatest part of the Salt-water is congeal'd or kern'd; or till a fresh Supply of it comes in again from the Sea. This Water is known to come in only at that one Passage on the N.

part

part of the Pond; where also it is deepest. It was at a Spring of the *New* Moon when I was there; and I was told that it comes in at no other time but at the New Moon Spring-tides: but why that should be I can't guess. They who come hither to lade Salt rake it up as it Kerns, and lay it in heaps on the dry Land, before the Water breaks in anew: And this is observable of this Salt-Pond, that the Salt kerns only in the Dry Season, contrary to the Salt-Ponds in the *West-Indies*, particularly those of the Island *Salt-Tortuga*, which I have formerly mentioned [Vol. I. p. 56.] for they never Kern there till the Rains come in about *April*; and continue to do so in *May*, *June*, *July*, &c. while the Wet Season lasts; and not without some good Shower of Rain first: But the Reason also of this Difference between the Salt-Ponds of *Mayo*, and those of the *West-Indies*, why these should Kern in the Wet Season, and the former in the Dry Season, I shall leave to Philosophers.

Our Nation drive here a great Trade for Salt, and have commonly a Man of War here for the Guard of our Ships and Barks that come to take it in; of which I have been inform'd that in some Years there have not been less than 100 in a Year. It costs nothing but Men's Labour to rake it together, and wheel it out of the Pond,

An. 1699. except the Carriage: And that also is very cheap; the Inhabitants having plenty of Asses, for which they have little to do besides carrying the Salt from the Ponds to the Sea-side at the Season when Ships are here. The Inhabitants lade and drive their Asses themselves, being very glad to be imploy'd; for they have scarce any other Trade but this to get a Penny by. The Pond is not above half a Mile from the Landing-place, so that the Asses make a great many Trips in a day. They have a set number of Turns to and fro both Forenoon and Afternoon, which their Owners will not exceed. At the Landing-place there lies a *Frape*-boat, as our Seamen call it, to take in the *Salt*. 'Tis made purposely for this use, with a Deck reaching from the Stern a third part of the Boat; where there is a kind of Bulk-head that rises, not from the Boats bottom, but from the Edge of the Deck, to about 2 foot in heighth; all calk'd very tight. The Use of it is to keep the Waves from dashing into the Boat, when it lies with its Head to the Shore, to take in Salt: For here commonly runs a great Sea; and when the Boat lies so with its Head to the Shore, the Sea breaks in over the Stern, and would soon fill it, was it not for this Bulk-head, which stops the Waves that come flowing upon the Deck,

and

and makes them run off into the Sea on each side. To keep the Boat thus with the Head to the Shore, and the Stern to the Sea, there are two strong Stantions set up in the Boat; the one at the Head, the other in the middle of it, against the Bulk-head, and a Foot higher than the Bulk-head. There is a large Notch cut in the top of each of these Stantions big enough for a small Hazer or Rope to lie in; one end of which is fasten'd to a Post ashore, and the other to a Grapling or Anchor lying a pretty way off at Sea: This Rope serveth to hale the Boat in and out, and the Stantions serve to keep her fast, so that she cannot swing to either side when the Rope is hal'd tight: For the Sea would else fill her, or toss her ashore and stave her. The better to prevent her staving and to keep her the tighter together, there are two sets of Ropes more: The first going athwart from Gunnal to Gunnal, which, when the Rowers Benches are laid, bind the Boats sides so hard against the Ends of the Benches that they cannot easily fall asunder, while the Benches and Ropes mutually help each other; the Ropes keeping the Boats sides from flying off, and the Benches from being crush'd together inwards. Of these Ropes there are usually but two, dividing the Boats length, as they go across the Sides, into there equal

An. 1699. parts. The other ſet of Ropes are more in number, and are ſo plac'd as to keep the Ribs and Planks of the Boat from ſtarting off. For this purpoſe there are holes made at certain diſtances through the Edge of the Keel that runs along on the inſide of the Boat; through which theſe Ropes paſſing are laid along the Ribs, ſo as to line them, or be themſelves as Ribs upon them, being made faſt to them by Rattan's brought thither, or ſmall Cords twiſted cloſe about both Ropes and Ribs, up to the Gunnal: By which means tho' ſeveral of the Nails or Pegs of the Boat ſhould by any ſhock fall out, yet the Ropes of theſe two ſets might hold her together: Eſpecially with the help of a Rope going quite round about the Gunnal on the out-ſide, as our Long-boats have. And ſuch is the Care taken to ſtrengthen the Boats; from which girding them with Ropes, which our Seamen call *Fraping*, they have the Name of Frape-boats. Two Men ſuffice to hale her in and out, and take in the Salt from Shore (which is brought in Bags) and put it out again. As ſoon as the Boat is brought nigh enough to the Shore, he who ſtands by the Bulk-head takes inſtantly a turn with the Hazer about the Bulk-head-Stantion; and that ſtops her faſt before the Sea can turn her aſide: And when the two Men have got in their Lading,

ing, they hale off to Sea, till they come a little without the ſwell; where they remove the Salt into another Boat that carries it on board the Ship. Without ſuch a *Frape*-boat here is but bad Landing at any time: for tho' 'tis commonly very ſmooth in the Road, yet there falls a great Sea on the Shore, ſo that every Ship that comes here ſhould have ſuch a *Boat*, and bring, or make, or borrow one of other Ships that happen to be here; for the Inhabitants have none. I have been thus particular in the Deſcription of theſe *Frape*-boats, becauſe of the Uſe they may be of in any Places where a great Sea falls in upon the Shore; as it doth eſpecially in many open Roads in the *East* and *West-Indies*; where they might therefore be very ſerviceable; but I never ſaw any of them there.

The Iſland *Mayo* is generally barren, being dry, as I ſaid; and the beſt of it is but a very indifferent Soil. The ſandy Bank that pens in the Salt-pond hath a ſort of Silk Cotton growing upon it, and a Plant that runs along upon the Ground, branching out like a Vine, but with thick broad Leaves. The Silk-Cotton grows on tender Shrubs, 3 or 4 Foot high, in Cods as big as an Apple, but of a long ſhape; which when ripe open at one end, parting leiſurely into 4 quarters; and at the firſt open-

An. 1699. ing the Cotton breaks forth. It may be of use for stuffing of Pillows, or the like; but else is of no value, any more than that of the great Cotton-tree. I took of these Cods before they were quite ripe, and laid them in my Chest; and in two or three days they would open and throw out the Cotton. Others I have bound fast with Strings, so that the Cod could not open; and in a few Days after, as soon as I slackned the String never so little, the Cod would burst, and the Cotton fly out forceably, at a very little hole, just as the Pulp out of a roasting Apple, till all has been out of the Cod. I met with this sort of Cotton afterwards at *Timor* (where it was ripe in *November*) and no where else in all my Travels; but I found two other sorts of Silk-cotton at *Brazil*, which I shall there describe. The right Cotton-Shrub grows here also, but not on the Sand-bank. I saw some Bushes of it near the Shore; but the most of it is planted in the middle of the Isle, where the Inhabitants live, Cotton-cloth being their chief Manufacture; but neither is there any great store of this Cotton. There also are some Trees within the Island, but none to be seen near the Sea-side; nothing but a few Bushes scattering up and down against the sides of the adjacent Hills; for, as I said before, the Land is pretty high from

the

the Sea. The Soil is for the most part either a sort of Sand, or loose crumbling Stone, without any fresh Water Ponds or Streams, to moisten it; but only Showers in the Wet-season, which run off as fast as they fall: except a small Spring in the middle of the Isle, from which proceeds a little Stream of Water that runs through a Valley between the Hills. There the Inhabitants live in three small Towns, having a Church and Padre in each Town: And these Towns, as I was inform'd, are 6 or 7 miles from the Road. *Pinose* is said to be the chief Town, and to have two Churches: St. *Johns* the next; and the third *Lagoa*. The Houses are very mean; small, low Things. They build with Fig-tree; here being, as I was told, no other Trees fit to build with. The Rafters are a sort of wild Cane. The Fruits of this Isle are chiefly Figs, and Water-Melons. They have also *Callavances* (a sort of Pulse like *French* Beans) and Pumpkins, for ordinary Food. The Fowls are Flamingo's, Great Curlews, and *Guinea*-Hens; which the Natives of those Islands call *Gallena Pintada*, or the Painted Hen; but in *Jamaica*, where I have seen also those Birds in the dry Savannah's and Woods, (for they love to run about in such Places) they are call'd *Guinea*-Hens. They seem to be much of the Nature of Partridges.

 They

An. 1699. They are bigger than our Hens, have long Legs, and will run apace. They can fly too, but not far, having large heavy Bodies, and but short Wings, and short Tails: As I have generally observ'd that Birds have seldom long Tails unless such as fly much; in which their Tails are usually serviceable to their turning about, as a Rudder to a Ship or Boat. These Birds have thick and strong, yet sharp Bills, pretty long Claws, and short Tails. They feed on the Ground, either on Worms, which they find by tearing open the Earth; or on Grashoppers, which are plentiful here. The Feathers of these Birds are speckled with dark and light Gray; the Spots so regular and uniform, that they look more beautiful than many Birds that are deck'd with gayer Feathers. Their Necks are small and long; their Heads also but little. The Cocks have a small rising on their Crowns, like a sort of a Comb. 'Tis of the colour of a dry Wall-Nut-shell, and very hard. They have a small red Gill on each side of their Heads, like Ears, strutting out downwards; but the Hens have none. They are so strong that one cannot hold them; and very hardy. They are very good Meat, tender, and sweet; and in some the Flesh is extraordinary white; tho' some others have black Flesh: but both sorts are very good.

good. The Natives take them with Dogs, running them down whenever they please; for here are abundance of them. You shall see 2 or 300 in a company. I had several brought aboard alive, where they throve very well; some of them 16 or 18 Months; when they began to pine. When they are taken young they will become tame like our Hens. The *Flamingo*'s I have already describ'd at large, [Vol. I. p. 79.] They have also many other sort of Fowls, *viz.* Pidgeons and Turtle-doves; *Miniota*'s, a sort of Land-fowls as big as Crows, of a grey colour, and good Food; *Crusia*'s, another sort of grey-colour'd Fowl almost as big as a Crow, which are only seen in the Night (probably a sort of Owls) and are said to be good for consumptive People, but eaten by none else. *Rabek*'s, a sort of large grey eatable Fowls with long Necks and Legs, not unlike Herons; and many kinds of small Birds.

Of Land-Animals, here are Goats, as I said formerly, and Asses good store. When I was here before they were said to have had a great many Bulls and Cows: But the Pirates, who have since miserably infested all these Islands, have much lessen'd the number of those; not having spar'd the Inhabitants themselves: for at my being there this time the

Gover-

An. 1699. Governor of *Mayo* was but newly return'd from being a Prisoner among them, they having taken him away, and carried him about with them for a Year or two.

The Sea is plentifully stock'd with Fish of divers sorts, *viz.* Dolphins, Boneta's, Mullets, Snappers, Silver-fish, Garfish, *&c.* and here is a good Bay to hale a Sain or Net in. I hal'd mine several times, and to good purpose; dragging ashore at one time 6 dozen of great Fish, most of them large Mullets of a foot and a half or two foot long. Here are also Porposes, and a small sort of Whales, that commonly visit this Road every day. I have already said, [Vol. I. p. 75.] That the Months of *May, June, July* and *August,* (that is, the Wet Season) are the time, when the Green Turtle come hither, and go ashore to lay their Eggs. I look upon it as a thing worth taking Notice of, that the Turtle should always, both in North and South Latitude, lay their Eggs in the Wet Months. It might be thought, considering what great Rains there are then in some places where these Creatures lay, that their Eggs should be spoiled by them. But the Rain, tho' violent, is soon soaked up by the Sand, wherein the Eggs are buried; and perhaps sinks not so deep into it as the Eggs are laid: And keeping down the Heat may make the Sand hotter below

than

than it was before, like a Hot-bed. What- An. 1699.
ever the Reaſon may be why Providence
determines theſe Creatures to this Seaſon
of laying their Eggs, rather than the Dry,
in Fact it is ſo, as I have conſtantly obſerv'd; and that not only with the Sea-
Turtle, but with all other ſorts of Amphibious Animals that lay Eggs; as Croco-
dils, Alligator's, Guano's, *&c.* The Inhabitants of this Iſland, even their Governour and *Padre*'s, are all Negro's, Wool-
pated like their *African*-Neighbours; from
whom 'tis like they are deſcended; tho'
being Subjects to the *Portugueſe* they have
their Religion and Language. They are
ſtout, luſty, well-limb'd People, both Men
and Women, fat and fleſhy; and they
and their Children as round and plump
as little Porpoſes; tho' the Iſland appears
ſo barren to a Stranger as ſcarce to have
Food for its Inhabitants. I inquired how
many People there might be on the Iſle;
And was told by one of the *Padre*'s, that
here were 230 Souls in all. The Negro-
Governor has his Patent from the *Portugueſe* Governor of St. *Jago.* He is a very
civil and ſenſible poor Man; and they are
generally a good ſort of People. He ex-
pects a ſmall Preſent from every Commander that lades Salt here; and is glad to
be invited aboard their Ships. He ſpends
moſt of his time with the *Engliſh* in the

An. 1699. Salting Season, which is his Harvest; and indeed, all the Islanders are then fully employed in getting somewhat; for they have no Vessels of their own to Trade with, nor do any *Portuguese*-Vessels come hither: scarce any but *English*, on whom they depend for Trade; and tho' Subjects of *Portugal*, have a particular Value for us. We don't pay them for their Salt, but for the Labour of themselves and their Beasts in lading it: for which we give them Victuals, some Mony, and old Cloaths, *viz.* Hats, Shirts, and other Cloaths: by which means many of them are indifferently well rigg'd; but some of them go almost Naked. When the Turtle-season comes in they watch the Sandy-bays in the Night, to turn them; and having small Huts at particular Places on the Bays to keep them from the Rain, and to sleep in: And this is another Harvest they have for Food; for by Report there come a great many Turtle to this and the rest of the *Cape Verd Islands*. When the Turtle Season is over they have little to do but to hunt for *Guinea*-Hens, and manage their small Plantations. But by these means they have all the Year some Employment or other; whereby they get a Subsistence, tho' but little else. When any of them are desirous to go over to St. *Jago* they get a Licence from the Governor, and desire passage in any

any *English* Ship that is going thither:
And indeed all Ships that lade Salt here will be obliged to touch at St. *Jago* for Water, for here at the Bay is none, not so much as for Drinking. 'Tis true there is a small Well of brackish Water not half a mile from the Landing-place, which the Asses that carry Salt drink at; but 'tis very bad Water. Asses themselves are a Commodity in some of these Islands, several of our Ships coming hither purposely to freight with them, and carry them to *Barbadoes* and our other Plantations. I stay'd at *Mayo* 6 days, and got 7 or 8 Tun of Salt aboard for my Voyage: In which time there came also into this Road several Sail of Merchants Ships for Salt; all bound with it for *Newfoundland.*

The 19th day of *February*, at about One a Clock in the Morning I weighed from *Mayo*-Road, in order to Water at St. *Jago*, which was about 5 or 6 Leagues to the Westward. We coasted along the Island *St. Jago*, and past by the Port on the East of it, I mention'd formerly [Vol. I. p. 76.] which they call *Praya*; where some English outward-bound *East-India* Men still touch, but not so many of them as heretofore. We saw the Fort upon the Hill, the Houses and Coco-nut Trees: But I would not go in to anchor here, because I expected better Water on the S. W. of the Island,

An. 1699. Island, at St. *Jago* Town. By 8 a Clock in the Morning we saw the Ships in that Road, being within 3 Leagues of it: But were forc'd to keep Turning many hours to get in, the Flaws of Wind coming so uncertain; as they do especially to the *Leeward* of Islands that are High Land. At length two *Portuguese* boats came off to help tow us in; and about 3 a Clock in the Afternoon we came to an Anchor; and took the Prospect of the Town, [Table II. Nº. 5.] We found here, besides two *Portuguese*-Ships bound for *Brazil*, whose Boats had tow'd us in; an *English* Pink that had taken in Asses at one of the *Cape Verd* Islands, and was bound to *Barbadoes* with them. Next Morning I went Ashore with my Officers to the Governor, who treated us with Sweet-meats: I told him, the occasion of my coming was chiefly for Water; and that I desired also to take in some Refreshments of Fowls, *&c.* He said I was welcom, and that he would order the Townsmen to bring their Commodities to a certain House, where I might purchase what I had occasion for: I told him I had not Mony, but would exchange some of the Salt which I brought from *Mayo* for their Commodities. He reply'd, that Salt was indeed an acceptable Commodity with the poor People, but that if I design'd to

buy

buy any Cattle, I must give Mony for them. I contented my self with taking in Dunghil Fowls: The Governor ordering a Cryer to go about the Town and give notice to the People, that they might repair to such a place with Fowls, and Maiz for feeding them, where they might get Salt in exchange for them: So I sent on board for Salt, and ordered some of my Men to truck the same for the Fowls and Maiz, while the rest of them were busie in filling of Water. This is the effect of their keeping no Boats of their own on the several Islands, that they are glad to buy even their own Salt of Foreigners, for want of being able to transport it themselves from Island to Island.

An. 1699.

St. Jago Town lies on the S. W. part of the Island, in Lat. about 15 Deg. N. and is the Seat of the General Governour, and of the Bishop of all the *Cape Verd* Islands. This Town stands scattering against the sides of two Mountains, between which there is a deep Valley, which is about 200 Yards wide against the Sea; but within a quarter of a mile it closes up so as not to be 40 Yards wide. In the Valley, by the Sea, there is a stragling Street, Houses on each side, and a Run of Water in the bottom, which empties it self into a fine small Cove or sandy Bay, where the Sea is commonly very smooth:

so

An.1699. ſo that here is good Wat'ring and good Landing at any time; tho' the Road be rocky and bad for Ships. Juſt by the Landing-place there is a ſmall Fort, almoſt level with the Sea, where is always a Court of Guard kept. On the top of the Hill, above the Town, there is another Fort; which, by the Wall that is to be ſeen from the Road, ſeems to be a large Place. They have Canon mounted there, but how many I know not: Neither what uſe that Fort can be of, except it be for Salutes. The Town may conſiſt of 2 or 300 Houſes, all built of rough Stone; having alſo one Convent, and one Church.

The People in general are black, or at leaſt of a mixt colour, except only ſome few of the better ſort, *viz.* the Governor, the Biſhop, ſome Gentlemen, and ſome of the Padres; for ſome of theſe alſo are black. The People about *Praya* are Thieviſh; but theſe of *St. Jago* Town, living under their Governour's Eye, are more orderly; tho' generally poor, having little Trade: Yet beſides chance Ships of other Nations, there come hither a *Portugueſe* Ship or two every Year, in their way to *Brazil.* Theſe vend among them a few *European* Commodies, and take of their principal Manufactures, *viz.* ſtriped Cotton-cloth, which they carry with them to *Brazil.* Here is alſo another Ship comes hither from *Portugal*

tugal for Sugar, their other Manufacture, and returns with it directly thither: For 'tis reported that there are several small Sugar-works on this Island, from which they send home near 100 Tun every year; and they have plenty of Cotton growing up in the Country, wherewith they cloath themselves, and send also a great deal to *Brazil*. They have Vines, of which they make some Wine: but the *European* Ships furnish them with better; tho' they drink but little of any. Their chief Fruits are, (besides Plantains in abundance) Oranges, Lemons, Citrons, Melons, (both Musk and Water-melons) Limes, Guava's, Pomgranates, Quinces, Custard-Apples, and Papah's, *&c.*

An. 1699.

The Custard-Apple (as we call it) is a Fruit as big as a *Pomegranate*, and much of the same colour. The out-side Husk, Shell or Rind, is for substance and thickness between the Shell of a Pomegranate, and the Peel of a *Sevil*-Orange; softer than this, yet more brittle than that. The Coat or Covering is also remarkable in that it is beset round with small regular Knobs or Risings; and the inside of the Fruit is full of a white soft Pulp, sweet and very pleasant, and most resembling a Custard of any thing, both in Colour and Tast: From whence probably it is called a Custard-Apple by our *English*. It has in the mid-

An. 1699. dle a few small black Stones or Kernels; but no Core, for 'tis all Pulp. The Tree that bears this Fruit is about the bigness of a Quince-tree, with long, small, and thick-set Branches spread much abroad: At the Extremity of here and there one of which the Fruit grows upon a Stalk of its own about 9 or 10 Inches long, slender and tough, and hanging down with its own weight. A large Tree of this sort does not bear usually above 20 or 30 Apples; seldom more. This Fruit grows in most Countries within the *Tropicks*. I have seen of them (tho' I omitted the Description of them before) all over the *West-Indies*, both Continent and Islands; as also in *Brazil*, and in the *East-Indies*.

The *Papah* too is found in all these Countries, though I have not hitherto describ'd it. It is a Fruit about the bigness of a Musk-Melon, hollow as that is, and much resembling it in Shape and Colour, both outside and inside: Only in the middle, instead of flat Kernels, which the Melons have, these have a handful of small blackish Seeds, about the bigness of Pepper-corns; whose Taste is also hot on the Tongue somewhat like Pepper. The Fruit it self is sweet, soft and luscious, when ripe; but while green 'tis hard and unsavory: tho' even then being boiled and eaten with Salt-pork or Beef, it serves in-

stead

ſtead of Turnips, and is as much eſteemed. An. 1683.
The Papah-Tree is about 10 or 12 Foot high. The Body near the Ground may be a Foot and an half or 2 Foot Diameter; and it grows up tapering to the top. It has no Branches at all, but only large Leaves growing immediately upon Stalks from the Body. The Leaves are of a roundiſh Form and jag'd about the Edges, having their Stalks or Stumps longer or ſhorter as they grow near or further from the top. They begin to ſpring from out of the Body of the Tree at about 6 or 7 Foot heighth from the Ground, the Trunk being bare below: but above that the Leaves grow thicker and larger ſtill towards its Top, where they are cloſe and broad. The Fruit grows only among the Leaves; and thickeſt among the thickeſt of them; inſomuch that towards the top of the Tree the *Papah*'s ſprings forth from its Body as thick as they can ſtick one by another. But then lower down, where the Leaves are thinner, the Fruit is larger, and of the ſize I have deſcrib'd: And at the Top, where they are thick, they are but ſmall, and no bigger than ordinary Turnips; yet taſted like the reſt.

Their chief Land-Animals are their Bullocks, which are ſaid to be many; tho' they askt us 20 Dollars apiece for them: They have alſo Horſes, Aſſes, and

*An.*1699. Mules, Deer, Goats, Hogs, and black-fac'd long-tail'd Monkeys. Of Fowls they have Cocks and Hens, Ducks, *Guinea*-Hens, both tame and wild, Parakites, Parrots, Pidgeons, Turtle-Doves, Herons, Hawks, Crab-catchers, Galdens, (a larger sort of Crab-catchers) Curlew's, &c. Their Fish is the same as at *Mayo* and the rest of these Islands, and for the most part these Islands have the same Beasts and Birds also: But some of the Isles have Pasturage and Employment for some particular Beasts more than other; and the Birds are incourag'd, by Woods for shelter, and Maiz and Fruits for Food, to flock rather to some of the Islands (as to this of St. *Jago*) than to others.

St. *Jago* Road is one of the worst that I have been in. There is not clean Ground enough for above 3 Ships; and those also must lie very near each other. One even of these must lie close to the Shore, with a Land-fast there: And that is the best for a small Ship. I should not have come in here if I had not been told that it was a good secure Place; but I found it so much otherways, that I was in pain to be gone. Captain *Barefoot*, who came to an Anchor while I was here, in foul Ground, lost quickly 2 Anchors; and I had lost a

small

ſmall one. The Iſland *Fogo* ſhews its ſelf *An.* 1699. from this Road very plain, at about 7 or 8 Leagues diſtance; and in the Night we ſaw the Flames of Fire iſſuing from its Top.

CHAP. II.

The A.'s Deliberation on the Sequel of his Voyage, and Departure from St. Jago. *His Course, and the Winds,* &c. *in crossing the* Line. *He stands away for the* Bay of All Saints *in* Brazil; *and why. His Arrival on that Coast and in the* Bay. *Of the several Forts, the Road, Situation, Town, and Buildings of* Bahia. *Of its Governor, Ships and Merchants; and Commodities to and from* Europe. *Claying of Sugar. The Season for the* European *Ships, and* Coire *Cables: Of their* Guinea-*trade, and of the Coasting-trade, and Whale-killing. Of the Inhabitants of* Bahia; *their carrying in Hammocks: their Artificers, Crane for Goods, and* Negro-*Slaves. Of the Country about* Bahia, *its Soil and Product. Its Timber-trees; the* Sapi-

An. 1699.

Amphisbæna, *small Black and small Grey-Snake; the great Land, and the great Water-Snake: and of the VVater-dog. Of their Sea-fish and Turtle; and of St.* Paul's-*Town.*

HAving dispatch'd my small Affairs at the *C. Verd* Islands, I meditated on the process of my Voyage. I thought it requisite to touch once more at a cultivated Place in these Seas, where my Men might be refresh'd, and might have a Market wherein to furnish themselves with Necessaries: For designing that my next Stretch should be quite to *N. Holland*, and knowing that after so long a Run nothing was to be expected there but fresh Water, if I could meet even with that there, I resolved upon putting in first at some Port of *Brazil*, and to provide my Self there with whatever I might have further Occasion for. Beside the refreshing and furnishing my Men, I aim'd also at the inuring them gradually and by intervals to the Fatigues that were to be expected in the remainder of the Voyage, which was to be in a part of the World they were altogether Strangers to; none of them, except two young Men, having ever crost the *Line*.

With

With this Design I sail'd from *St. Jago* on the 22d of *February*, with the Winds at E. N. E. and N. E. fair Weather, and a brisk Gale. We steered away S. S. E. and S. S. E. half East, till in the Lat. of 7 deg. 50 min. we met with many Riplings in the Sea like a Tide or strong Current, which setting against the Wind caus'd such a Ripling. We continu'd to meet these Currents from that Lat. till we came into the Lat. of 3 deg. 22 N. when they ceased. During this time we saw some Boneta's, and Sharks; catching one of these. We had the true General Trade-Wind blowing fresh at N. E. till in the Lat. of 4 deg. 40 min. N. when the Wind varied, and we had small Gales, with some Tornadoes. We were then to the East of *St. Jago* 4 deg. 54 min. when we got into Lat. 3 deg. 2 min. N. (where I said the Ripling ceas'd) and Long. to the East of *St. Jago* 5 deg. 2 min. we had the Wind whiffling between the S. by E. and E. by N. small Gales, frequent Calms, very black Clouds, with much Rain. In the Lat. of 3 deg. 8 min. N. and Long. E. from *St. Jago* 5 deg. 8 min. we had the Wind from the S. S. E. to the N. N. E. faint, and often interrupted with Calms. While we had Calms we had the opportunity of trying the Current we had met with hitherto, and found that it set N. E.

An. 1699.

by

An. 1699. by E. half a Knot, which is 12 mile in 24 hours: So that here it ran at the Rate of half a mile an hour, and had been much stronger before. The Rains held us by intervals till the Lat. of 1 deg. 0 min. N. with small Gales of Wind between S. S. E. and S. E. by E. and sometimes calm: Afterwards we had the Wind between the S. & S. S. E. till we crost the Line, small Winds, Calms, and pretty fair Weather. We saw but few Fish beside Porposes; but of them a great many, and struck one of them.

It was the 10th day of *March*, about the time of the *Equinox*, when we crost the *Equator*, having had all along from the Lat. of 4 deg. 40 min. N. where the True Trade-Wind left us, a great swell out of the S. E. and but small uncertain Gales, mostly Southerly, so that we crept to the Southward but slowly. I kept up against these as well as I could to the Southward, and when we had now and then a flurry of Wind at E. I still went away due South, purposely to get to the Southward as fast as I could; for while near the *Line* I expected to have but uncertain Winds, frequent Calms, Rains, Tornadoes, *&c.* which would not only retard my Course, but endanger Sickness also among my Men: especially those who were ill provided with Cloaths, or were too lazy to shift themselves

ſelves when they were drench'd with the Rains. The Heat of the Weather made them careleſs of doing this; but taking a Dram of Brandy, which I gave them when wet, with a Charge to ſhift themſelves, they would however lie down in their Hammocks with their Wet Cloaths; ſo that when they turn'd out they caus'd an ill ſmell where-ever they came, and their Hammocks would ſtink ſufficiently: that I think the Remedying of this is worth the Care of Commanders that croſs the *Line*; eſpecially when they are, it may be, a Month or more e'er they get out of the Rains, at ſome times of the Year, as in *June*, *July*, or *Auguſt*.

What I have here ſaid about the Currents, Winds, Calms, *&c.* in this Paſſage is chiefly for the farther Illuſtration of what I have heretofore obſerv'd in general about theſe Matters, and eſpecially as to Croſſing the Line, in my *Diſcourſe of the Winds*, &c. *in the Torrid Zone*: [See Vol. II. Part 3. p. 5, 6.] Which Obſervations I have had very much confirm'd to me in the Courſe of this Voyage; and I ſhall particularize in ſeveral of the chief of them as they come in my Way. And indeed I think I may ſay this of the Main of the Obſervations in that *Treatiſe*, that the clear Satisfaction I had about them, and how much I might rely upon them, was a great Eaſe to my Mind

An. 1699. Mind during this Vexatious Voyage; wherein the Ignorance, and Obstinacy withal, of some under me, occasion'd me a great deal of Tronble: tho' they found all along, and were often forc'd to acknowledge it, that I was seldom out in my Conjectures, when I told them usually beforehand what Winds, *&c.* we should meet with at such or such particular Places we should come at.

Pernambuc was the Port that I designed for at my first setting out from *St. Jago*; it being a Place most proper for my purpose, by reason of its Situation, lying near the Extremity of *C. St. Augustine*, the Easternmost Promontory of *Brazil*; by which means it not only enjoys the greater benefit of the Sea-breezes, and is consequently more healthy than other Places to the Southward, but is withal less subject to the Southerly Coasting-Trade-winds, that blow half the Year on this Shore; which were now drawing on, and might be troublesome to me: So that I might both hope to reach soonest *Pernambuc*, as most directly and nearest in my Run; and might thence also more easily get away to the Southward than from *Bahia de Todos los Santos*, or *Ria Janeira*.

But notwithstanding these Advantages I propos'd to my self in going to *Pernambuc*, I was soon put by that Design through the

re-

refractoriness of some under me, and the Discontents and Backwardness of some of my Men. For the Calms and Shiftings of Wind which I met with, as I was to expect, in crossing the Line, made them, who were unacquainted with these Matters, almost heartless as to the persuit of the Voyage, as thinking we should never be able to weather C. *St. Augustine*: And though I told them that by that time we should get to about three Degrees South of the Line, we should again have a True brisk General Trade-Wind from the North East, that would carry us to what part of *Brazil* we pleas'd, yet they would not believe it till they found it so. This, with some other unforeseen Accidents, not necessary to be mention'd in this place, meeting with the Aversion of my Men to a long unknown Voyage, made me justly apprehensive of their Revolting, and was a great Trouble and Hindrance to me. So that I was obliged partly to alter my Measures, and met with many Difficulties, the Particulars of which I shall not trouble the Reader with: But I mention thus much of it in general for my own necessary Vindication,

An. 1699. tion, in my taking ſuch Meaſures ſometimes for proſecuting the Voyage as the ſtate of my Ships Crew, rather than my own Judgment and Experience, determin'd me to. The Diſorders of my Ship made me think at preſent that *Pernambuc* would not be ſo fit a Place for me; being told that Ships ride there two or three Leagues from the Town, under the Command of no Forts; ſo that whenever I ſhould have been aſhore it might have been eaſy for my diſcontented Crew to have cut or ſlipt their Cables, and have gone away from me: Many of them diſcovering already an Intention to return to *England*, and ſome of them declaring openly that they would go no further onwards than *Brazil*. I alter'd my Courſe therefore, and ſtood away for *Bahio de todos los Santos*, or the *Bay of All Saints*, where I hop'd to have the Governor's help, if need ſhould require, for ſecuring my Ship from any ſuch Mutinous Attempt; being forc'd to keep my ſelf all the way upon my Guard, and to lie with my Officers, ſuch as I could truſt, and with ſmall Arms, upon the Quarter-deck; it ſcarce being ſafe for me to lie in my Cabbin, by Reaſon of the Diſcontents among my Men.

On

On the 23d of *March* we saw the Land *An.*1699.
of *Brazil*; having had thither, from the time when we came into the True Trade-Wind again after crossing the Line, very fair Weather and brisk Gales, mostly at E. N. E. The Land we saw was about 20 Leagues to the North of *Bahia*; so I coasted along Shore to the Southward. This Coast is rather low than high, with Sandy-Bays all along by the Sea.

A little within Land are many very white Spots of Sand, appearing like Snow; and the Coast looks very pleasant, being checker'd with Woods and Savanahs. The Trees in general are not tall; but they are green and flourishing. There are many small Houses by the Sea-side, whose Inhabitants are chiefly Fishermen. They come off to Sea on Barklogs, made of several Logs fasten'd side to side, that have one or two Masts with Sails to them. There are two Men in each Barklog, one at either end, having small low Benches, raised a little above the Logs, to sit and fish on, and two Baskets hanging up at the Mast or Masts; one to put their Provisions in, the other for their Fish. Many of these were a Fishing now, and two of them came aboard, of whom I bought some Fish. In the Afternoon we sailed by one very remarkable piece of Land, where, on a small pleasant Hill, there was a

Church

An. 1699. Church dedicated to the Virgin *Mary*. See a Sight of some parts of this Coast [Table III. N°. 1, 2, 3, 4, 5.] and of the Hill the Church stands on [Table III. N°. 1.]

I coasted along till the Evening, and then brought to, and lay by till the next Morning. About 2 hours after we were brought to, there came a Sail out of the Offin (from Seaward) and lay by about a Mile to Windward of us, and so lay all Night. In the Morning, upon speaking with her, she proved to be a *Portuguese* Ship bound to *Bahia*; therefore I sent my Boat aboard and desired to have one of his Mates to Pilot me in. He answer'd, That he had not a Mate capable of it, but that he would sail in before me, and shew me the way; and that if he went in to the Harbour in the Night, he would hang out a Light for me. He said we had not far in and might reach it before Night with a tolerable Gale; but that with so small an one as now we had we could not do it: So we jog'd on till Night, and then he accordingly hung out his Light, which we steered after, sounding as we went in. I kept all my Men on Deck, and had an Anchor ready to let go on occasion. We had the Tide of Ebb against us, so that we went in but slowly; and it was about the middle of the Night when we anchor'd. Immediately the *Portuguese* Master came aboard

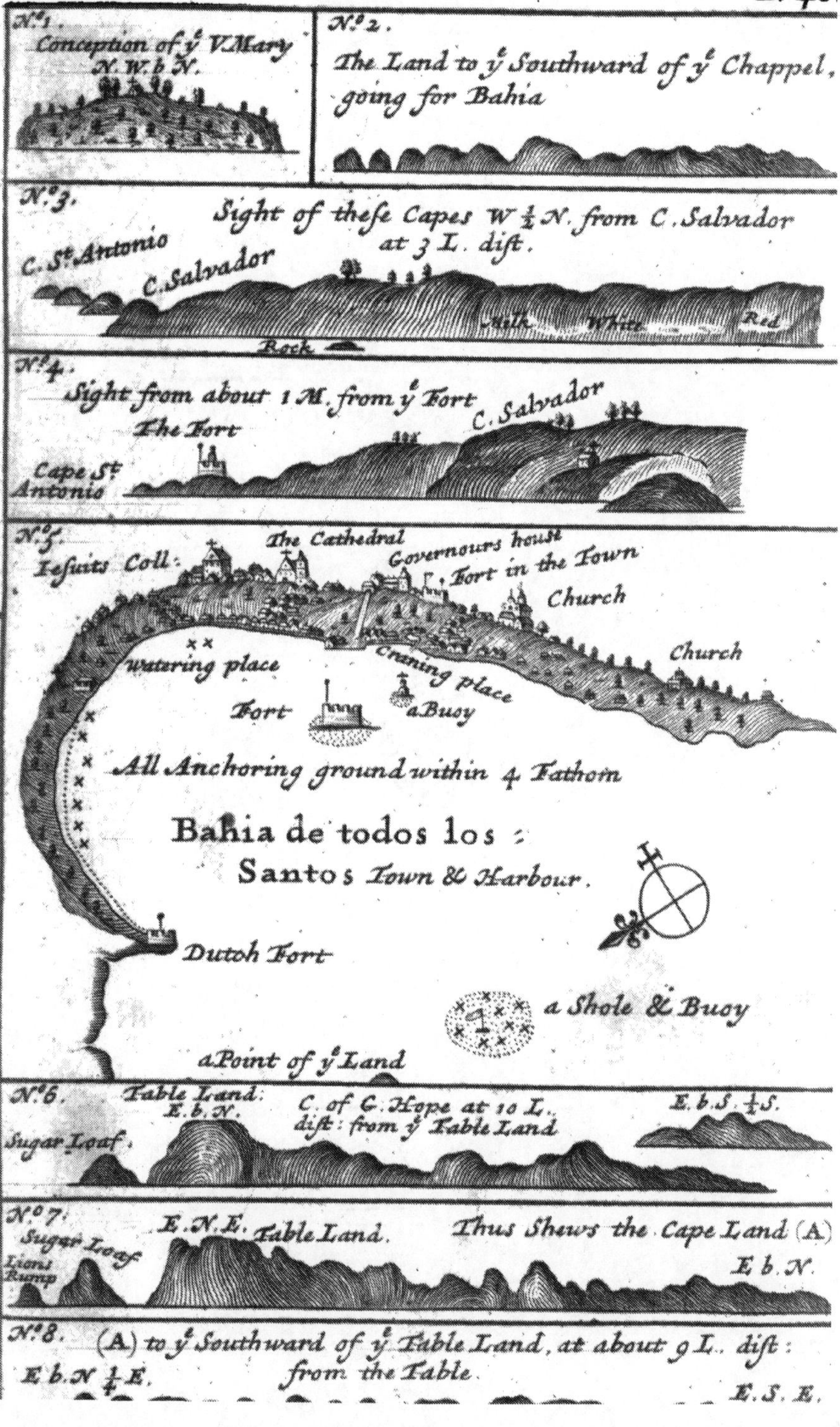

N.º 1.
Conception of yᵉ V. Mary N. W. b N.
N.º 2.
The Land to yᵉ Southward of yᵉ Chappel, going for Bahia
N.º 3.
Sight of these Capes W ½ N. from C. Salvador at 3 L. dist.
C. Sᵗ Antonio
C. Salvador
Milk
White
Red
Rock
N.º 4.
Sight from about 1 M. from yᵉ Fort
The Fort
C. Salvador
Cape Sᵗ Antonio
N.º 5.
Iesuits Coll.
The Cathedral
Governours house
Fort in the Town
Church
Church
Watering place
Craning Place
Fort
a Buoy
All Anchoring ground within 4 Fathom
Bahia de todos los Santos Town & Harbour.
Dutch Fort
a Shole & Buoy
a Point of yᵉ Land
N.º 6.
Table Land: E. b. N.
C. of G. Hope at 10 L. dist: from yᵉ Table Land
E. b. S. ½ S.
Sugar Loaf.
N.º 7.
Sugar Loaf
E. N. E.
Table Land.
Thus Shews the Cape Land (A)
Lions Rump
E b. N.
N.º 8.
(A) to yᵉ Southward of yᵉ Table Land, at about 9 L. dist: from the Table
E b. N ½ E.
E. S. E.

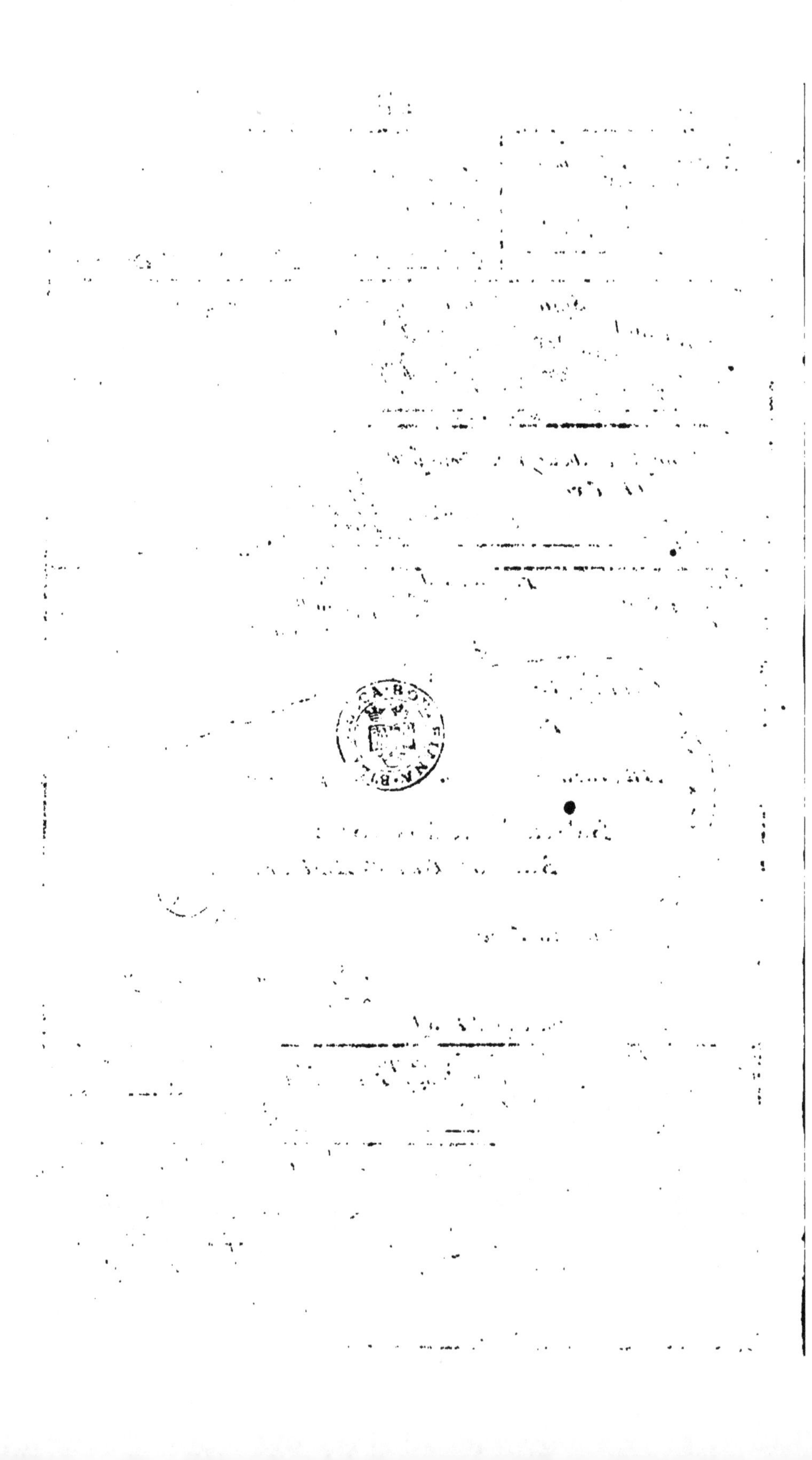

aboard to ſee me, to whom I returned Thanks for his Civilities; and indeed I found much Reſpect, not only from this Gentleman, but from all of that Nation both here and in other Places, who were ready to ſerve me on all Occaſions. The Place that we anchored in was about two Miles from the Harbour where the Ships generally ride; but the Fear I had leſt my People ſhould run away with the Ship made me haſten to get a Licence from the Governor, to run up into the Harbour, and ride among their Ships, cloſe by one of their Forts. So on the 25th of *March* about 10 a Clock in the Morning the Tide ſerving I went thither, being Piloted by the Super-intendant there, whoſe Buſineſs it is to carry up all the King of *Portugal's* Ships that come hither, and to ſee them well moored. He brought us to an Anchor right againſt the Town, at the outer part of the Harbour, which was then full of Ships, within 150 yards of a ſmall Fort that ſtands on a Rock half a mile from the Shore. See a Proſpect of the Harbour and the Town, as it appear'd to us while we lay at Anchor, [Table III. N°. 5.]

Bahia de todos los Santos lies in Lat. 13 deg. S. It is the moſt conſiderable Town in *Brazil*, whether in reſpect of the Beauty of its Buildings, its Bulk, or its Trade and Revenue. It has the convenience of a

E good

An. 1699. good Harbour that is capable of receiving Ships of the greatest Burthen: The Entrance of which is guarded with a strong Fort standing without the Harbour, call'd *St. Antonio:* A Sight of which I have given [Table III. N°. 4.] as it appear'd to us the Afternoon before we came in; and its Lights (which they hang out purposely for Ships) we saw the same Night. There are other smaller Forts that command the Harbour, one of which stands on a Rock in the Sea, about half a mile from the Shore. Close by this Fort all Ships must pass that anchor here, and must ride also within half a mile of it at farthest between this and another Fort (that stands on a Point at the inner part of the Harbour and is called the *Dutch* Fort) but must ride nearest to the former, all along against the Town: where there is good holding Ground, and less exposed to the Southerly Winds that blow very hard here. They commonly set in about *April*, but blow hardest in *May*, *June*, *July* and *August*: but the Place where the Ships ride is exposed to these Winds not above 3 Points of the Compass.

Beside these there is another Fort fronting the Harbour, and standing on the Hill upon which the Town stands. The Town it self consists of about 2000 Houses; the major part of which cannot be seen from

from the Harbour: but so many as appear in sight, with a great mixture of Trees between them, and all placed on a rising Hill, make a very pleasant Prospect; as may be judg'd by the Draught, [Table III. N°. 5.]

An. 1699.

There are in the Town 13 Churches, Chapels, Hospitals, Convents, beside one Nunnery; *viz.* the *Ecclesia Major* or Cathedral, the Jesuits College, which are the chief, and both in sight from the Harbour: *St. Antonio*, *Sta. Barbara*, both Parish-Churches; the *Franciscans* Church, and the *Dominicans*; and two Convents of *Carmelites*; a Chapel for Seamen close by the Sea-side, where Boats commonly land, and the Seamen go immediately to Prayers; another Chapel for poor People, at the farther end of the same Street, which runs along by the Shore; and a third Chapel for Soldiers, at the edge of the Town, remote from the Sea; and an Hospital in the middle of the Town. The Nunnery stands at the outer-edge of the Town next the Fields, wherein by Report there are 70 Nuns. Here lives an Archbishop who has a fine Palace in the Town; and the Governor's Palace is a fair Stone-building, and looks handsome to the Sea, tho' but indifferently furnish'd within: both *Spaniards* and *Portuguese* in their Plantations abroad, as I have generally

An. 1699. obſerv'd, affecting to have large Houſes; but are little curious about Furniture, except Pictures ſome of them. The Houſes of the Town are 2 or 3 Stories high, the Walls thick and ſtrong, being built with Stone, with a Covering of Pantile; and many of them have Balconies. The principal Streets are large, and all of them pav'd or pitch'd with ſmall Stones. There are alſo Parades in the moſt eminent Places of the Town, and many Gardens, as well within the Town as in the Out-parts of it, wherein are Fruit-trees, Herbs, Salladings and Flowers in great variety, but order'd with no great Care nor Art.

The Governor who reſides here is call'd *Don John de Lancaſtario*, being deſcended, as they ſay, from our *Engliſh Lancaſter* Family; and he has a reſpect for our Nation on that account, calling them his Countrymen. I waited on him ſeveral times and always found him very courteous and civil. Here are about 400 Soldiers in Gariſon. They commonly draw up and exerciſe in a large Parade before the Governor's Houſe; and many of them attend him when he goes abroad. The Soldiers are decently clad in brown Linen, which in theſe hot Countries is far better than Woollen; but I never ſaw any clad in Linen but only theſe. Beſide the Soldiers in Pay, he can ſoon have ſome thouſands of Men up in

Arms

Arms on occasion. The Magazine is on the Skirts of the Town, on a small rising between the Nunnery and the Soldiers Church. 'Tis big enough to hold 2 or 3000 Barrels of Powder; but I was told it seldom has more than 100, sometimes but 80. There are always a Band of Soldiers to guard it, and Centinels looking out both Day and Night.

An. 1699.

A great many Merchants always reside at *Bahia*; for 'tis a Place of great Trade: I found here above 30 great Ships from *Europe*, with two of the King of *Portugal's* Ships of War for their Convoy; beside two Ships that Traded to *Africa* only, either to *Angola*, *Gamba*, or other Places on the Coast of *Guinea*; and abundance of small Craft, that only run to and fro on this Coast, carrying Commodities from one part of *Brazil* to another.

The Merchants that live here are said to be Rich, and to have many *Negro* Slaves in their Houses, both of Men and Women. Themselves are chiefly *Portuguese*, Foreigners having but little Commerce with them; yet here was one Mr. *Cock* an *English* Merchant, a very civil Gentleman and of good Repute. He had a Patent to be our *English* Consul, but did not care to take upon him any Publick Character, because *English* Ships seldom come hither, here having been none in 11

An. 1699. or 12 years before this time. Here was also a *Dane*, and a *French* Merchant or two; but all have their Effects transported to and from *Europe* in *Portuguese* Ships, none of any other Nation being admitted to Trade hither. There is a Custom-house by the Sea-side, where all Goods imported or exported are entred. And to prevent Abuses there are 5 or 6 Boats that take their turns to row about the Harbour, searching any Boats they suspect to be running of Goods.

The chief Commodities that the *European* Ships bring hither, are Linnen-cloaths, both course and fine; some Woollens also, as Bays, Searges, Perpetuana's, *&c.* Hats, Stockings, both of Silk and Thread, Bisket-bread, Wheat-flower, Wine (chiefly *Port*) Oil-Olive, Butter, Cheese, *&c.* and Salt-beef and Pork would there also be good Commodities. They bring hither also Iron, and all sorts of Iron-Tools; Pewter-Vessels of all sorts, as Dishes, Plates, Spoons, *&c.* Looking-glasses, Beads, and other Toys; and the Ships that touch at *St. Jago* bring thence, as I said, Cotton cloath, which is afterwards sent to *Angola*.

The *European* Ships carry from hence Sugar, Tobacco, either in Roll or Snuff, never in Leaf, that I know of: These are the Staple Commodities. Besides which, here

here are Dye-woods, as Fustick, *&c.* with Woods for other uses, as speckled Wood, *Brazil*, &c. They also carry home raw Hides, Tallow, Train-Oil of Whales, *&c.* Here are also kept tame Monkeys, Parrots, Parakites, *&c.* which the Seamen carry home.

An. 1699.

The Sugar of this Country is much better than that which we bring home from our Plantations: for all the Sugar that is made here is clay'd, which makes it whiter and finer than our *Muscovada*, as we call our unrefin'd Sugar. Our Planters seldom refine any with Clay, unless sometimes a little to send home as Presents for their Friends in *England*. Their way of doing it is by taking some of the whitest Clay and mixing it with Water, 'till 'tis like Cream. With this they fill up the Pans of Sugar, that are sunk 2 or 3 Inches below the Brim by the draining of the Molosses out of it: First scraping off the thin hard Crust of the Sugar that lies at the top, and would hinder the Water of the Clay from soaking through the Sugar of the Pan. The refining is made by this Percolation. For 10 or 12 days time that the Clayish Liquor lies soaking down the Pan, the white Water whitens the Sugar as it passes thro' it; and the gross Body of the Clay it self grows hard on the top, and may be taken off at pleasure; when scraping off with a

*An.*1699. Knife the very upper part of the Sugar, which will be a little sullied, that which is underneath will be White almost to the bottom: and such as is called *Brazil* Sugar is thus Whitened. When I was here this Sugar was sold for 50 *s. per* 100 ℔. and the Bottoms of the Pots, which is very course Sugar, for about 20 *s. per* 100 ℔. both sorts being then scarce; for here was not enough to lade the Ships, and therefore some of them were to lie here till the next Season.

The *European* Ships commonly arrive here in *February* or *March*, and they have generally quick Passages; finding at that time of the Year brisk Gales to bring them to the Line, little Trouble, then, in crossing it, and brisk E. N. E. Winds afterwards to bring them hither. They commonly return from hence about the latter end of *May*, or in *June*. 'Twas said when I was here that the Ships would sail hence the 20th day of *May*; and therefore they were all very busy, some in taking in their Goods, others in Careening and making themselves ready. The Ships that come hither usually Careen at their first coming; here being a Hulk belonging to the King for that purpose. This Hulk is under the charge of the Superintendent I spoke of, who has a certain Sum of Mony for every Ship that Careens by her. He also pro-

vides

vides Firing and other Necessaries for that purpose: and the Ships do commonly hire of the Merchants here each 2 Cables to moor by all the time they lie here, and so save their own Hempen Cables; for these are made of a sort of Hair, that grows on a certain kind of Trees, hanging down from the Top of their Bodies, and is very like the black *Coyre* in the *East-Indies*, if not the same. These Cables are strong and lasting: And so much for the *European* Ships.

An. 1699.

The Ships that use the *Guinea*-Trade are small Vessels in comparison of the former. They carry out from hence Rum, Sugar, the Cotton-cloaths of St. *Jago*, Beads, *&c.* and bring in return, Gold, Ivory, and Slaves; making very good returns.

The small Craft that belong to this Town are chiefly imployed in carrying *European* Goods from *Bahia*, the Center of the *Brasilian* Trade, to the other Places on this Coast; bringing back hither Sugar, Tobacco, *&c.* They are sailed chiefly with Negro-Slaves; and about *Christmas* these are mostly imployed in Whale-killing: for about that time of the Year a sort Whales, as they call them, are very thick on this Coast. They come in also into the Harbours and inland Lakes, where the Seamen go out and kill them. The Fat of

them

An. 1699. them is boyled to Oyl; the Lean is eaten by the Slaves and poor People: And I was told by one that had frequently eaten of it that the Flesh was very sweet and wholesome. These are said to be but small Whales: yet here are so many, and so easily kill'd, that they get a great deal of Mony by it. Those that strike them buy their Licence for it of the King: And I was informed that he receives 30000 Dollars *per Annum* for this Fishery. All the small Vessels that use this Coasting Traffick are built here; and so are some Men of War also for the King's Service. There was one a Building when I was here, a Ship of 40 or 50 Guns: And the Timber of this Country is very good and proper for this purpose. I was told it was very strong, and more durable than any we have in *Europe*: and they have enough of it. As for their Ships that use the *European* Trade, some of them that I saw there were *English* built, taken from us by the *French* during the late War, and sold by them to the *Portugese*.

Besides Merchants and others that Trade by Sea from this Port, here are other pretty Wealthy Men, and several Artificers and Trades-men of most sorts, who by Labour and Industry maintain themselves very well; especially such as can arrive at the purchase of a *Negro*-Slave or two.

And

And indeed, excepting People of the lowest degree of all, here are ſcarce any but what keep Slaves in their Houſes. The Richer Sort, beſides the Slaves of both Sexes whom they keep for ſervile Uſes in their Houſes, have Men-ſlaves who wait on them abroad, for State; either running by their Horſes-ſides when they ride out, or to carry them to and fro on their Shoulders in the Town when they make ſhort Viſits near home. Every Gentleman or Merchant is provided with Things neceſſary for this ſort of Carriage. The main Thing is a pretty large Cotton Hammock of the *West-India* Faſhion, but moſtly dyed Blue, with large Fringes of the ſame, hanging down on each ſide. This is carry'd on the *Negro*'s Shoulders by the help of a Bambo about 12 or 14 Foot long, to which the Hammock is hung; and a Covering comes over the Pole, hanging down on each ſide like a Curtain: So that the Perſon ſo carry'd cannot be ſeen unleſs he pleaſes; but may either ly down, having Pillows for his Head; or may ſit up by being a little ſupported with theſe Pillows, and by letting both his Legs hang out over one ſide of the Hammock. When he hath a mind to be ſeen he puts by his Curtain, and ſalutes every one of his Acquaintance whom he meets in the Streets; for they take a piece of Pride in greeting

one

An. 1699. one another from their Hammocks, and will hold long Conferences thus in the Streets: But then their two Slaves who carry the Hammock have each a ſtrong well-made Staff, with a fine Iron Fork at the upper end, and a ſharp Iron below, like the Reſt for a Musket, which they ſtick faſt in the Ground, and let the Pole or Bambo of the Hammock reſt upon them, till their Maſters Buſineſs or the Complement is over. There is ſcarce a Man of any faſhion, eſpecially a Woman, will paſs the Streets but ſo carried in a Hammock. The chief Mechanick Traders here, are Smiths, Hatters, Shoemakers, Tanners, Sawyers, Carpenters, Coopers, *&c.* Here are alſo Taylors, Butchers, *&c.* which laſt kill the Bullocks very dexterouſly, ſticking them at one Blow with a ſharp-pointed Knife in the Nape of the Neck, having firſt drawn them cloſe to a Rail; but they dreſs them very ſlovenly. It being *Lent* when I came hither there was no buying any Fleſh till *Eaſter*-Eve, when a great number of Bullocks were kill'd at once in the Slaughter-houſes within the Town, Men, Women and Children flocking thither with great Joy to buy, and a multitude of Dogs, almoſt ſtarv'd, following them; for whom the Meat ſeem'd fitteſt, it was ſo Lean: All theſe Trades-men buy *Negro's*, and train them

them up to their ſeveral Imployments, *An.1699.* which is a great help to them: and they having ſo frequent Trade to *Angola*, and other parts of *Guinea*, they have a conſtant ſupply of Blacks both for their Plantations and Town. Theſe Slaves are very uſeful in this Place for Carriage, as Porters; for as here is a great Trade by Sea, and the Landing-place is at the foot of a Hill, too ſteep for drawing with Carts, ſo there is great need of Slaves to carry Goods up into the Town, eſpecially for the inferiour ſort: but the Merchants have alſo the Convenience of a great Crane that goes with Ropes or Pullees, one end of which goes up while the other goes down. The Houſe in which this Crane is ſtands on the Brow of the Hill towards the Sea, hanging over the Precipice: and there are Planks ſet ſhelving againſt the Bank from thence to the Bottom, againſt which the Goods lean or ſlide as they are hoiſted up or let down. The *Negro*-Slaves in this Town are ſo numerous, that they make up the greateſt part or bulk of the Inhabitants: Every Houſe, as I ſaid, having ſome, both Men and Women, of them. Many of the *Portugueſe*, who are Batchelors, keep of theſe black Women for Miſſes, tho' they know the danger they are in of being poyſon'd by them, if ever they give them any occaſion of Jealouſy. A Gentleman of my Ac-

quaintance,

An. 1699. quaintance, who had been familiar with his Cook-maid, lay under ſome ſuch Apprehenſions from her when I was there. Theſe Slaves alſo of either Sex will eaſily be engaged to do any ſort of Miſchief; even to Murder, if they are hired to do it, eſpecially in the Night: for which Reaſon, I kept my Men on board as much as I could; for one of the *French* King's Ships being here had ſeveral Men murder'd by them in the Night, as I was credibly inform'd.

Having given this account of the Town of *Bahia*, I ſhall next ſay ſomewhat of the Country. There is a Salt-water Lake runs 40 Leagues, as I was told, up the Country, N. W. from the Sea, leaving the Town and *Dutch* Fort on the Starboard ſide. The Country all round about is for the moſt part a pretty flat even Ground, not high, nor yet very low: It is well watered with Rivers, Brooks and Springs, neither wants it for good Harbours, Navigable Creeks, and good Bays for Ships to ride in. The Soil in general is good, naturally producing very large Trees of divers ſorts, and fit for any uſes. The Savannahs alſo are loaden with Graſs, Herbs, and many ſorts of ſmaller Vegetables; and being cultivated, produce any thing that is proper for thoſe hot Countrys, as Sugar-Canes, Cotton, Indi-

co,

co, Tobacco, Maiz, Fruit-Trees of seve- *An.*1699. ral kinds, and Eatable Roots of all sorts. Of the several kinds of Trees that are here, I shall give an account of some, as I had it partly from an Inhabitant of *Bahia*, and partly from my knowledge of them otherwise, *viz. Sapiera*, *Vermiatico*, *Comesserie*, *Guitteba*, *Serrie*, as they were pronounc'd to me, three sorts of *Mangrove*, speckled Wood, Fustick, Cotton-Trees of three sorts, *&c.* together with Fruit-Trees of divers sorts that grow wild, beside such as are planted.

Of Timber-Trees, the *Sapiera* is said to be large and tall; it is very good Timber, and is made use of in building of Houses; so is the *Vermiatico*, a tall streight-bodied Tree, of which they make Plank 2 Foot broad, and they also make Canoa's with it. *Comesserie* and *Guitteba* are chiefly used in building Ships; these are as much esteemed here, as Oaks are in *England*, and they say either sort is harder and more durable than Oak. The *Serrie* is a sort of Tree much like Elm, very durable in water. Here are also all the three sorts of *Mangrove* Trees, *viz.* the Red, the White, and the Black, which I have described [*Vol.* I. *p.* 54.] The Bark of the Red Mangrove, is here used for Tanning of Leather, and they have great Tan-pits for it. The Black Mangrove grows lar-

ger

An. 1699. ger here than in the *West-Indies*, and of it they make good Plank. The White Mangrove is larger and tougher than in the *West-Indies*; of these they make Masts and Yards for Barks.

There grow here Wild or Bastard Coco Nut Trees, neither so large nor so tall as the common ones in the *East* or *West-Indies*. They bear Nuts as the others, but not a quarter so big as the right Coco-Nuts. The shell is full of Kernel, without any hollow Place or Water in it; and the Kernel is sweet and wholesome, but very hard both for the Teeth and for Digestion. These Nuts are in much esteem for making Beads for *Pater-noster*'s, Boles of Tobacco-pipes, and other Toys: and every small Shop here has a great many of them to sell. At the top of these Bastard Coco-trees, among the Branches, there grows a sort of long black Thread like Horse-hair, but much longer, which by the *Portuguese* is called *Tresabo*. Of this they make Cables which are very serviceable, strong and lasting; for they will not rot as Cables made of Hemp, tho' they ly exposed both to Wet and Heat. These are the Cables which I said they keep in their Harbours here, to let to hire to *European* Ships, and resemble the *Coyre*-Cables.

Here are three sorts of Cotton Trees that bear Silk-cotton. One sort is such as I have

have formerly described, [Vol. I. p. 165.] An. 1699. by the Name of the Cotton-tree. The other two sorts I never saw any where but here. The Trees of these latter sorts are but small in comparison of the former, which are reckon'd the biggest in all the *West-India* Woods; yet are however of a good bigness and heighth. One of these last sorts is not so full of Branches as the other of them; neither do they produce their Fruit the same time of the Year: for one sort had its Fruit just ripe, and was shedding its Leaves while the other sort was yet green, and its Fruit small and growing, having but newly done blossoming; the Tree being as full of young Fruit as an Apple-Tree ordinarily in *England*. These last yield very large Pods, about 6 Inches long, and as big as a Man's Arm. It is ripe in *September* and *October*; then the Pod opens, and the Cotton bursts out in a great Lump as big as a Man's Head. They gather these Pods before they open: otherways it would fly all away. It opens as well after 'tis gathered; and then they take out the Cotton, and preserve it to fill Pillows and Bolsters, for which use 'tis very much esteemed: but 'tis fit for nothing else, being so short that it cannot be spun. 'Tis of a tawney Colour; and the Seeds are black, very round, and as big as a white Pea. The other sort is ripe

An. 1699. in *March* or *April*. The Fruit or Pod is like a large Apple, and very round. The out-ſide Shell is as thick as the top of ones Finger. Within this there is a very thin whitiſh Bag or Skin which incloſeth the Cotton. When the Cotton-Apple is ripe the outer thick green Shell ſplits it ſelf into 5 equal parts from Stemb to Tail, and drops off, leaving the Cotton hanging upon the Stemb, only pent up in its fine Bag. A day or two afterwards the Cotton ſwells by the heat of the Sun, breaks the Bag and burſts out, as big as a Man's Head: And then as the Wind blows 'tis by degrees driven away, a little at a time, out of the Bag that ſtill hangs upon the Stemb, and is ſcatter'd about the Fields; the Bag ſoon following the Cotton, and the Stemb the Bag. Here is alſo a little of the right *Weſt-India* Cotton Shrub; but none of the Cotton is exported, nor do they make much Cloth of it.

This Country produces great variety of fine Fruits, as very good Oranges of 3 or 4 ſorts; (eſpecially one ſort of *China* Oranges;) Limes in abundance, Pomgranets, Pomecitrons, Plantains, Bonano's, right Coco-nuts, Guava's, Coco-plumbs, (call'd here *Munſberow*'s) Wild-Grapes, ſuch as I have deſcrib'd [Vol. II. Part 2. p. 46.] beſide ſuch Grapes as grow in *Europe*. Here are alſo Hog-plumbs, Cuſtard-Apples,

ples, *Sour-sops*, *Cashews*, *Papah's* (called here *Mamoons*) *Jennipah's* (called here *Jenni-papah's*) Manchineel-Apples and Mango's. Mango's are yet but rare here: I saw none of them but in the *Jesuit's* Garden, which has a great many fine Fruits, and some Cinamon-trees. These, both of them, were first brought from the *East-Indies*, and they thrive here very well: So do Pumplemusses, brought also from thence; and both *China* and *Sevil* Oranges are here very plentiful as well as good.

The *Sour-sop* (as we call it) is a large Fruit as big as a Man's Head, of a long or oval Shape, and of a green Colour; but one side is Yellowish when ripe. The outside Rind or Coat is pretty thick, and very rough, with small sharp Knobs; the inside is full of spungy Pulp, within which also are many black Seeds or Kernels, in shape and bigness like a Pumkin-seed. The Pulp is very juicy, of a pleasant Taste, and wholesome. You suck the Juice out of the Pulp, and so spit it out. The Tree or Shrub that bears this Fruit grows about 10 or 12 Foot high, with a small short Body; the Branches growing pretty strait up; for I did never see any of them spread abroad. The Twigs are slender and tough; and so is the Stemb of the Fruit. This Fruit grows also both in the *East* and *West-Indies*.

An. 1699. The *Cashew* is a Fruit as big as a Pippin, pretty long, and bigger near the Stemb than at the other end, growing tapering. The Rind is smooth and thin, of a red and yellow Colour. The Seed of this Fruit grows at the end of it; 'tis of an Olive Colour shaped like a Bean, and about the same bigness, but not altogether so flat. The Tree is as big as an Apple-tree, with Branches not thick, yet spreading off. The Boughs are gross, the Leaves broad and round, and in substance pretty thick. This Fruit is soft and spongy when ripe, and so full of Juice that in biting it the Juice will run out on both sides of ones Mouths. It is very pleasant, and gratefully rough on the Tongue; and is accounted a very wholesome Fruit. This grows both in the *East* and *West Indies*, where I have seen and eaten of it.

The *Jenipah* or *Jenipapah* is a sort of Fruit of the Calabash or Gourd-kind. It is about the bigness of a Duck-Egg, and somewhat of an Oval Shape; and is of a grey Colour. The Shell is not altogether so thick nor hard as a Calabash: 'Tis full of whitish Pulp mixt with small flat Seeds; and both Pulp and Seeds must be taken into the Mouth, where sucking out the Pulp you spit out Seeds. It is of a sharp and pleasing Taste, and is very innocent. The Tree that bears it is much like an Ash,

strait

strait-bodied, and of a good heighth; clean from Limbs till near the top, where there branches forth a small Head. The Rind is of a pale grey, and so is the Fruit. We us'd of this Tree to make Helves or Handles for Axes (for which it is very proper) in the Bay of *Campeachy*; where I have seen of them, and no where else but here.

Beside these, here are many sorts of Fruits which I have not met with any where but here; as *Arisah*'s, *Mericasah*'s, *Petango*'s, *&c.* *Arisah*'s are an excellent Fruit, not much bigger than a large Cherry; shaped like a Catherine-Pear, being small at the Stemb, and swelling bigger towards the end. They are of a greenish colour, and have small Seeds as big as Mustard-Seeds. They are somewhat tart, yet pleasant, and very wholsom, and may be eaten by sick People.

Mericasah's, are an excellent Fruit, of which there are two sorts; one growing on a small Tree or Shrub, which is counted the best; the other growing on a kind of Shrub like a Vine, which they plant about Arbours to make a shade, having many broad Leaves. The Fruit is as big as a small Orange, round and green. When they are ripe they are soft and fit to eat; full of white pulp mixt thick with little black Seeds, and there is no separating one from the other, till they are in your

An. 1699. Mouth; when you suck in the white Pulp and spit out the Stones. They are tart, pleasant, and very wholsome.

Petango's are a small red Fruit, that grow also on small Trees, and are as big as Cherries, but not so Globular, having one flat side, and also 5 or 6 small protulerant Ridges. 'Tis a very pleasant tart Fruit, and has a pretty large flattish Stone in the middle.

Petumbo's, are a yellow Fruit (growing on a shrub like a Vine) bigger than Cherries, with a pretty large Stone: These are sweet, but rough in the Mouth.

Mungaroo's, are a Fruit as big as Cherries, red on one side and white on the other side: They are said to be full of small Seeds, which are commonly swallowed in eating them.

Muckishaw's, are said to be a Fruit as big as Crab-Apples, growing on large Trees. They have also small Seeds in the middle, and are well tasted.

Ingwa's, are a Fruit like the Locust-Fruit, 4 Inches long, and one broad. They grow on high Trees.

Otee, is a Fruit as big as a large Coco-Nut. It hath a Husk on the outside, and a large Stone within, and is a accounted a very fine Fruit.

Musteran-

Musteran-de-ova's, are a round Fruit as big as large Hazel-Nuts, cover'd with thin brittle shells of a blackish colour: They have a small Stone in the middle, inclosed within a black pulpy substance, which is of a pleasant taste. The outside shell is chewed with the Fruit, and spit out with the Stone, when the pulp is suck'd from them. The Tree that bears this Fruit is tall, large, and very hard Wood. I have not seen any of these five last named Fruits, but had them thus described to me by an *Irish* Inhabitant of *Bahia*; tho' as to this last, I am apt to believe, I may have both seen and eaten of them in *Achin* in *Sumatra*.

Palm-Berries (called here *Dendees*) grow plentifully about *Bahia*; the largest are as big as Wall-nuts; they grow in bunches on the top of the Body of the Tree, among the Roots of the Branches or Leaves, as all Fruits of the Palm kind do. These are the same kind of Berries or Nuts as those they make the Palm-Oyl with on the Coast of *Guinea*, where they abound: And I was told that they make Oyl with them here also. They sometimes roast and eat them; but when I had one roasted to prove it, I did not like it.

Physick-Nuts, as our Seamen call them, are called here *Pineon*; and *Agnus Castus* is called here *Carrepat*: These both grow here:

An. 1699. here: so do *Mendibees*, a Fruit like *Physick-Nuts*. They scorch them in a Pan over the fire before they eat them.

Here are also great plenty of Cabbage-Trees, and other Fruits, which I did not get information about, and which I had not the opportunity of seeing; because this was not the Season, it being our Spring, and consequently their Autumn, when their best Fruits were gone, tho' some were left. However I saw abundance of wild Berries in the Woods and Field, but I could not learn their Names or Nature.

They have withal good plenty of ground Fruit, as *Callavances*, Pine-Apples, Pumkins, Water-Melons, Musk-Melons, Cucumbers; and Roots, as Yams, Potato's Cassava's, *&c.* Garden Herbs also good store; as Cabbages, Turnips, Onions, Leeks, and abundance of other Salading, and for the Pot. Drugs of several sorts, *viz.* Sassafras, Snake-Root, *&c.* Beside the Woods I mentioned for Dying, and other Uses, as Fustick, Speckled-wood, *&c.*

I brought home with me from hence a good number of Plants, dried between the leaves of Books; of some of the choicest of which, that are not spoil'd, I may give a Specimen at the *End* of the *Book*.

Here are said to be great plenty and variety of Wild-Fowl, *viz.* *Yemma's*, *Maccaw's* (which are called here *Jackoo's*, and are a larger sort of Parrots, and scarcer) Parrots,

rots, Parakites, Flamingo's, Carrion-Crows, Chattering-Crows, Cockrecoes, Bill-Birds finely painted, Corresoes, Doves, Pigeons, *Jenetees*, Clocking-Hens, Crabcatchers, Galdens, Currecoo's, Moscovy Ducks, common Ducks, Widgeons, Teal, Curlews, Men of War Birds, Booby's, Noddy's, Pelicans, *&c.*

An. 1699.

The *Yemma* is bigger than a Swan, grey-feathered, with a long thick sharp-pointed Bill.

The Carrion-Crow and Chattering-Crows, are called here *Mackeraw*'s, and are like those I described in the *West-Indies*, [*Vol.* II. *Part* II. *p.* 67.] The Bill of the Chattering-Crow is black, and the Upper-Bill is round, bending downwards like a Hawks-Bill, rising up in a ridge almost semicircular, and very sharp, both at the Ridge or Convexity, and at the Point or Extremity: The Lower-Bill is flat and shuts even with it. I was told by a *Portegueze* here, that their *Negro*-Wenches make Love-Potions with these Birds. And the *Portuguese* care not to let them have any of these Birds, to keep them from that Superstition: As I found one Afternoon when I was in the Fields with a Padre and another, who shot two of them, and hid them, as they said, for that reason. They are not good Food, but their Bills are reckoned a good Antidote against Poison. The

An. 1699. The *Bill-Birds* are so called by the *English*, from their monstrous Bills, which are as big as their Bodies. I saw none of these Birds here, but saw several of the Breasts flea'd off and dried, for the beauty of them; the Feathers were curiously colour'd with red, yellow, and Orange-colour.

The *Curreso*'s (called here *Mackeraw*'s) are such as are in the Bay of *Campeachy* [Vol. 2. Part 2. p. 67.]

Turtle-Doves are in great plenty here; and two sorts of Wild Pigeons; the one sort blackish, the other a light grey: The blackish or dark grey are the bigger, being as large as our Wood-Quests, or Wood-Pigeons in *England*. Both sorts are very good Meat; and are in such plenty from *May* till *September*, that a Man may shoot 8 or 10 Dozen in several Shots at one standing, in a close misty Morning, when they come to feed on Berries that grow in the Woods.

The *Jenetee* is a Bird as big as a Lark, with blackish Feathers, and yellow Legs and Feet. 'Tis accounted very wholsom Food.

Clocking-Hens, are much like the Crab-catchers, which I have described [*Vol.* II. *Part* 2. *p.* 70.] but the Legs are not altogether so long. They keep always in swampy wet places, tho' their Claws are like

like Land-Fowl's Claws. They make a Noiſe or *Cluck* like our Brood-Hens, or Dunghil-Hens, when they have Chickens, and for that reaſon they are called by the *Engliſh* Clocking Hens. There are many of them in the Bay of *Campeachy* (tho' I omitted to ſpeak of them there) and elſewhere in the *Weſt-Indies*. There are both here and there four ſorts of theſe long-leg'd Fowls, near a-kin to each other, as ſo many *Sub-Species* of the ſame Kind; *viz.* Crab catchers, Clocking-Hens, Galdens (which three are in ſhape and colour like Herons in *England*, but leſs; the *Galden*, the biggeſt of the three, the Crab-catcher the ſmalleſt;) and a fourth ſort which are black, but ſhaped like the other, having long Legs and ſhort Tails; theſe are about the bigneſs of *Crab-catchers*, and feed as they do.

Currecoos, are Water Fowls, as big as pretty large Chickens, of a bluiſh colour, with ſhort Legs and Tail; they feed alſo in ſwampy Ground, and are very good Meat. I have not ſeen of them elſewhere.

The Wild-Ducks here are ſaid to be of two ſorts, the *Muſcovy*, and the common-Ducks. In the wet Seaſon here are abundance of them, but in the dry time but few. Wigeon and Teal alſo are ſaid to be in great plenty here in the wet Seaſon.

To

An. 1699. To the Southward of *Bahia* there are also Ostridges in great plenty, tho', 'tis said, they are not so large as those of *Africa*: They are found chiefly in the Southern Parts of *Brasil*, especially among the large Savanahs near the River of *Plate*; and from thence further South towards the Streights of *Magellan*.

As for Tame Fowl at *Bahia*, the chief beside their Ducks, are Dunghil-Fowls, of which they have two sorts; one sort much of the size of our Cocks and Hens; the other very large; and the Feathers of these last are a long time coming forth; so that you see them very naked when half grown; but when they are full grown and well feathered, they appear very large Fowls, as indeed they are; neither do they want for price; for they are sold at *Bahia* for half a Crown or three Shillings apiece, just as they are brought first to Market out of the Countrey, when they are so lean as to be scarce fit to Eat.

The Land Animals here are Horses, black Cattle, Sheep, Goats, Rabbits, Hogs, Leopards, Tigers, Foxes, Monkeys, Pecary (a sort of wild Hogs, called here *Pica*) Armadillo, Alligaters, Guano's (called *Quittee*) Lizards, Serpents, Toads, Frogs, and a sort of Amphibious Creatures called by the *Portuguese Cachora's de agua*, in *English* Water-Dogs.

The

The Leopards and Tigers of this Country are said to be large and very fierce: But here on the Coast they are either destroyed, or driven back towards the heart of the Country; and therefore are seldom found but in the Borders and Out-plantations, where they oftentimes do Mischief. Here are three or four sorts of Monkeys, of different Sizes and Colours. One sort is very large; and another sort is very small: These last are ugly in Shape and Feature, and have a strong Scent of Musk.

An. 1699.

Here are several sorts of Serpents, many of them vastly great, and most of them very venomous: As the Rattle-snake for one: and for Venome, a small Green Snake is bad enough, no bigger than the Stemb of a Tobacco-pipe, and about 18 Inches long, very common here.

They have here also the *Amphisbæna*, or Two-headed Snake, of a grey Colour, mixt with blackish Stripes, whose Bite is reckon'd to be incurable. 'Tis said to be blind, tho' it has two small Specks in each Head like Eyes: but whether it sees or not I cannot tell. They say it lives like a Mole, mostly under Ground; and that when it is found above Ground it is easily kill'd, because it moves but slowly: Neither is its Sight (if it hath any) so good as to discern any one that comes near to Kill

it:

An. 1699. as few of these Creatures fly at a Man, or hurt him but when he comes in their Way. 'Tis about 14 Inches long, and about the bigness of the inner joint of a Man's middle Finger; being of one and the same bigness from one end to the other, with a Head at each end, (as they said; for I cannot vouch it, for one I had was cut short at one end) and both alike in shape and bigness; and 'tis said to move with either Head formost, indifferently; whence 'tis called by the *Portugueze*, *Cobra de dos Cabesas*, the Snake with two Heads.

The small black Snake is a very venomous Creature.

There is also a grey Snake, with red and brown Spots all over its Back. 'Tis as big as a Man's Arm, and about 3 Foot long, and is said to be venomous. I saw one of these.

Here are two sorts of very large Snakes or Serpents: One of 'em a Land-snake, the other a Water-snake. The Land-snake is of a grey colour, and about 18 or 20 Foot long: Not very Venomous, but Ravenous. I was promised the sight of one of their Skins, but wanted opportunity.

The Water-snake is said to be near 30 Foot long. These live wholly in the Water, either in large Rivers, or great Lakes, and prey upon any Creature that comes within

within their reach, be it Man or Beast. They draw their Prey to them with their Tails: for when they see any thing on the Banks of the River or Lake where they lurk, they swing about their Tails 10 or 12 Foot over the Bank; and whatever stands within their Sweep is snatcht with great Violence into the River, and drowned by them. Nay 'tis reported very credibly that if they see only a shade of any Animal at all on the Water, they will flourish their Tails to bring in the Man or Beast whose shade they see, and are oftentimes too successful in it. Wherefore Men that have Business near any place where these Water-Monsters are suspected to lurk, are always provided with a Gun, which they often fire, and that scares them away, or keeps them quiet. They are said to have great Heads, and strong Teeth about 6 Inches long. I was told by an *Irish* Man who lived here, that his Wives Father was very near being taken by one of them about the time of my first Arrival here, when his Father was with him up in the Country: for the Beast flourisht his Tail for him, but came not nigh enough by a yard or two; however it scared him sufficiently.

The Amphibious Creatures here which I said are called by the *Portuguese Cuchora's de Agua*, or Water-dogs, are said to be as big

An. 1699. big as small Mastives, and are all hairy and shaggy from Head to Tail. They have 4 short Legs, a pretty long Head and short Tail; and are of a blackish colour. They live in fresh Water-ponds, and oftentimes come ashore and Sun themselves; but retire to the Water if assaulted. They are eaten, and said to be good Food. Several of these Creatures which I have now spoken of I have not seen, but inform'd my self about them while I was here at *Bahia*, from sober and sensible Persons among the Inhabitants, among whom I met with some that could speak *English*.

In the Sea upon this Coast there is great store and diversity of Fish, *viz.* Jew-fish, for which there is a great Market at *Bahia* in *Lent*: Tarpom's, Mullets, Groopers, Snooks, Gar-fish (called here *Goolion*'s), *Gorasses*, Barrama's, Coquinda's, Cavallie's, Cuchora's (or Dog-fish) Conger-Eeles, Herrings (as I was told) the *Serrew*, the *Olio de Boy*, (I write and spell them just as they were named to me) Whales, &c.

Here is also Shell-fish (tho' in less plenty about *Bahia* than on other parts of the Coast) *viz.* Lobsters, Craw-fish, Shrimps, Crabs, Oysters of the common sort, Conchs, Wilks, Cockles, Muscles, Perriwinkles, &c. Here are three sorts of Sea-Turtle, *viz.* Hawksbill, Loggerhead, and

Green:

Green: but none of them are in any esteem, neither *Spaniards* nor *Portuguese* loving them: Nay they have a great Antipathy against them, and would much rather eat a Porpose, tho' our *English* count the Green-Turtle very extraordinary Food. The Reason that is commonly given in the *West-Indies* for the *Spaniards* not caring to eat of them, is the fear they have lest being usually foul-bodied, and many of them pox'd (lying, as they do, so promiscuously with their Negrines and other She-slaves) they should break out loathsomely like Lepers; which this sort of Food, 'tis said, does much incline Men to do, searching the Body, and driving out any such gross Humors: for which cause many of our *English* Valetudinarians have gone from *Jamaica* (tho' there they have also Turtle) to the I. *Caimanes*, at the Laying-time, to live wholly upon Turtle that then abound there; purposely to have their Bodies scour'd by this Food, and their Distempers driven out: and have been said to have found many of them good Success in it. But this by the way. The Hawksbill-Turtle on this Coast of *Brazil* is most sought after of any, for its Shell, which by Report of those I have convers'd with at *Bahia*, is the clearest and best-clouded Tortoise-shell in the World. I had some of it shewn me, which was indeed as good

An. 1699. as I ever saw. They get a pretty deal of it in some Parts on this Coast; but 'tis very dear.

Besides this Port of *Bahia de todos los Santos*, there are two more principal Ports on *Brazil*, where *European* Ships Trade, *viz. Pernambuc* and *Ria Janeira*; and I was told that there go as many Ships to each of these Places as to *Bahia*, and two Men of War to each Place for their Convoys. Of the other Ports in this Country none is of greater Note than that of St. *Paul's*, where they gather much Gold; but the Inhabitants are said to be a sort of *Banditti*, or loose People that live under no Government: but their Gold brings them all sorts of Commodities that they need, as Cloths, Arms, Ammunition, *&c.* The Town is said to be large and strong.

CHAP.

CHAP. III.

The A.'s Stay and Business at Bahia: *Of the Winds, and Seasons of the Year there. His departure for* N. Holland. C. Salvadore. *The Winds on the* Brasilian *Coast; and* Abrohlo *Sh*[illegible]; *Fish, and Birds: The Shear-water Bird, and Cooking of Sharks. Excessive number of Birds about a dead Whale; of the Pintado-Bird, and the Petrel,* &c. *Of a Bird that shews the* C. of G. Hope *to be near: of the Sea-reckonings, and* Variations: *and a* Table *of all the* Variations *observ'd in this Voyage. Occurrences near the* Cape; *and the A.'s passing by it. Of the Westerly Winds beyond it: A Storm, and its Presages. The A.'s Course to* N. Holland; *and Signs of approaching it. Another* Abrohlo *Shole and Storm, and the A.'s Arrival on part of* N. Holland. *That part describ'd; and* Shark's *Bay, where he first An-*

An. 1699.

chors. Of the Land there, Vegetables, Birds, &c. A particular sort of Guano: *Fish, and beautiful Shells; Turtle, large Shark, and Water-Serpents. The A.'s removing to another part of* N. Holland: *Dolphins, Whales, and more Sea-Serpents: and of a* Passage *or* Streight *suspected here: Of the Vegetables, Birds, and Fish. He anchors on a third Part of* N. Holland, *and digs Wells, but brackish. Of the Inhabitants there, the great Tides, the Vegetables and Animals,* &c.

MY stay here at *Bahia* was about a Month: during which time the Vice-Roy of *Goa* came hither from thence in a great Ship, said to be richly laden with all sorts of *India* Goods; but she did not break Bulk here, being bound home for *Lisbon*: only the Vice-Roy intended to refresh his Men (of whom he had lost many, and most of the rest were very sickly, having been 4 Months in their Voyage hither) and so to take in Water, and depart for *Europe*, in Company with the other *Por-*

Portuguese Ships thither Bound; who had Orders to be ready to Sail by the twentieth of *May*. He desir'd me to carry a Letter for him, directed to his Successor, the new Vice-Roy of *Goa*: Which I did; sending it thither afterwards by Captain *Hammond*, whom I found near the *Cape of Good Hope*. The Refreshing my Men, and taking in Water, was the main also of my Business here; beside the having the better opportunity to compose the Disorders among my Crew: Which, as I have before related, were grown to so great a Heighth, that they could not without great Difficulty be appeased: However, finding Opportunity, during my stay in this Place, to allay in some measure the Ferment that had been raised among my Men, I now set my self to provide for the carrying on of my Voyage with more Heart than before, and put all Hands to work, in order to it, as fast as the backwardness of my Men would permit; who shew'd continually their unwillingness to proceed farther. Besides, their Heads were generally fill'd with strange Notions of Southerly Winds that were now setting in (and there had been already some Flurries of them)

An. 1699. which, as they surmiz'd, would hinder any farther Attempts of going on to the Southward, so long as they should last.

The Winds begin to shift here in *April* and *September*, and the Seasons of the Year (the Dry and the Wet) alter with them. In *April* the Southerly Winds make their entrance on this Coast, bringing in the Wet Season, with violent Tornado's, Thunder and Lightning, and much Rain. In *September* the other Coasting Trade, at East North East comes in, and clears the Sky, bringing fair Weather. This, as to the change of Wind, is what I have observ'd Vol. II. Part 3. p. 19. but as to the change of Weather accompanying it so exactly here at *Bahia*, this is a particular Exception to what I have Experienc'd in all other Places of South Latitudes that I have been in between the *Tropicks*, or those I have heard of; for there the Dry Seasons sets in, in *April*, and the Wet about *October* or *November*, sooner or later (as I have said that they are, in South Latitudes, the Reverse of the Seasons, or Weather, in the same Months in N. Latitudes Vol. II. Part 3. p. 77.) whereas on this Coast of *Brazil*, the Wet Season

comes

comes in in *April*, at the same time that it doth in N. Latitudes, and the Dry (as I have said here) in *September*; the Rains here not lasting so far in the year as in other Places: for in *September* the Weather is usually so fair, that in the latter part of that Month they begin to cut their Sugar-Canes here, as I was told; for I enquired particularly about the Seasons: Though this, as to the Season of cutting of Canes, which I was now assur'd to be in *September*, agrees not very well with what I was formerly told [Vol. II. Part 3. p. 82.] that in *Brazil* they cut the Canes in *July*. And so, as to what is said a little lower in the same Page, that in managing their Canes they are not confin'd to the Seasons, this ought to have been express'd only of Planting them; for they never cut them but in the Dry Season.

But to return to the Southerly Winds, which came in (as I expected they would) while I was here: These daunted my Ship's Company very much, tho' I had told them they were to look for them: But they being ignorant as to what I told them farther, that these were only Coasting-Winds, sweeping the Shore to about 40 or 50 Leagues in breadth from it; and imagining that they had blown so all the Sea over, between *America* and *Africa*;

 and

*An.*1699. and being confirm'd in this their Opinion by the *Portugueſe* Pilots of the *European* Ships, with whom ſeveral of my Officers converſed much, and who were themſelves as ignorant that theſe were only Coaſting Trade-Winds (themſelves going away before them, in their return homewards, till they croſs the Line, and ſo having no experience of the Breadth of them) being thus poſſeſs'd with a Conceit that we could not Sail from hence till *September*; this made them ſtill the more remiſs in their Duties, and very liſtleſs to the getting Things in a readineſs for our Departure. However I was the more diligent my Self to have the Ship ſcrub'd, and to ſend my Water-Casks aſhore to get them trim'd, my Beer being now out. I went alſo to the Governor to get my Water fill'd; for here being but one Watering-place (and the Water running low, now at the end of the Dry Seaſon) it was always ſo crouded with the *European* Ships Boats, who were preparing to be gone, that my Men could ſeldom come nigh it, till the Governor very kindly ſent an Officer to clear the Water-place for my Men, and to ſtay there till my Wates-Casks were all full, whom I ſatisfied for his Pains. Here I alſo got Aboard 9 or 10 Tun of Ballaſt, and made my Boatſwain fit the

Rig-

Rigging that was amiſs: and I enquired *An.* 1699. alſo of my particular Officers whoſe Buſineſs it was, whether they wanted any Stores, eſpecially Pitch and Tar; for that here I wou'd ſupply my ſelf before I proceeded any farther: but they ſay they had enough, tho' it did not afterwards prove ſo.

I commonly went aſhore every day, either upon Buſineſs, or to recreate my ſelf in the Fields, which were very pleaſant, and the more for a ſhower of Rain now and then, that uſhers in the Wet Seaſon. Several ſorts of good Fruits were alſo ſtill remaining, eſpecially Oranges, which were in ſuch plenty, that I and all my Company ſtock'd our ſelves for our Voyage with them, and they did us a great kindneſs; and we took in alſo a good quantity of Rum and Sugar: But for Fowls, they being here lean and dear, I was glad I had ſtockt my ſelf at St. *Jago.* But by the little care my Officers took for freſh Proviſions, one might conclude, they did not think of going much farther. Beſides, I had like to have been imbroiled with the Clergy here (of the *Inquiſition*, as I ſuppoſe) and ſo my Voyage might have been hindred. What was ſaid to them of me, by ſome of my Company that went aſhore, I know not; but I was aſſur'd by a Merchant there, that if they got me into their Clutches (and

it

An. 1699. it ſeems, when I was laſt aſhore they had narrowly watch'd me) the Governor himſelf could not releaſe me. Beſides I might either be murther'd in the Streets, as he ſent me word, or Poyſoned, if I came aſhore any more; and therefore he adviſed me to ſtay aboard. Indeed I had now no further Buſineſs aſhore but to take leave of the Governor, and therefore took his Advice.

Our Stay here was till the 23d of *April.* I would have gone before if I could ſooner have fitted my ſelf; but was now earneſt to be gone, becauſe this Harbour lies open to the S. and S. S. W. which are raging Winds here, and now was the Seaſon for them. We had had two or three Touches of them; and one pretty ſevere: and the Ships ride there ſo near each other, that if a Cable ſhould fail, or an Anchor ſtart, you are inſtantly aboard of one Ship or other: and I was more afraid of being diſabled here in Harbour by theſe bluſtring Winds, than diſcouraged by them, as my People were, from proſecuting the Voyage; for at preſent I even wiſh'd for a brisk Southerly Wind as ſoon as I ſhould be once well out of the Harbour, to ſet me the ſooner into the True General Trade-Wind.

The Tide of Flood being ſpent, and having a fine Land-Breez on the 23d. in the

the Morning, I went away from the Anchoring-place before 'twas light; and then lay by till Day-light that we might see the better how to go out of the Harbour. I had a Pilot belonging to Mr. *Cock* who went out with me, to whom I gave three Dollars; but I found I could as well have gone out my Self, by the Soundings I made at coming in. The Wind was E. by N. and fair Weather. By 10 a Clock I was got past all danger, and then sent away my Pilot. At 12 Cape *Salvadore* bore N. distant 6 Leagues, and we had the Winds between the E. by N. and S. E. a considerable time, so that we kept along near the Shore, commonly in sight of it. The Southerly Blasts had now left us again; for they come at first in short Flurries, and shift to other Points (for 10 or 12 days sometimes) before they are quite set in: And we had uncertain Winds, between Sea and Land-Breezes, and the Coasting-Trade, which was its self unsetled.

An. 1699.

The Easterly-Winds at present made me doubt I should not weather a great Shoal which lies in Lat. between 18 deg. and 19 deg. S. and runs a great way into the Sea, directly from the Land, Easterly. Indeed the Weather was fair (and continued so a good while) so that I might the better avoid any Danger from it: and if the Wind came to the Southward I knew

I

An. 1699. I could ſtretch off to Sea; ſo that I jogg'd on couragiouſly. The 27th of *April* we ſaw a ſmall Brigantine under the Shore plying to the Southward. We alſo ſaw many Men of War-birds and Boobies, and abundance of *Albicore*-Fiſh. Having ſtill fair Weather, ſmall Gales, and ſome Calms, I had the opportunity of trying the Current, which I found to ſet ſometimes Northerly and ſometimes Southerly: and therefore knew I was ſtill within the Verge of the Tides. Being now in the Lat. of the *Abrohlo* Shoals, which I expected to meet with, I ſounded, and had Water leſſening from 40 to 33. and ſo to 25 Fathom: but then it roſe again to 33, 35, 37. &c. all Coral Rocks. Whilſt we were on this Shoal (which we croſt towards the further part of it from Land, where it lay deep, and ſo was not dangerous) we caught a great many Fiſh with Hook and Line; and by evening Amplitude we had 6 deg. 38 min. Eaſt Variation. This was the 27th of *April*; we were then in Lat. 18 deg. 13 min. S. and Eaſt Longitude from Cape *Salvadore* 31 min. On the 29th, being then in Lat. 18 deg. 39 min. S. we had ſmall Gales from the W. N. W. to the W. S. W. often ſhifting. The 30th we had the Winds from W. to S. S. E. Squals and Rain: and we ſaw ſome Dolphins and other Fiſh about us. We were

now

now out of sight of Land, and had been An. 1699.
so 4 or 5 Days: but the Wind's now hanging in the South was an apparent Sign that we were still too nigh the Shore to receive the True General East-Trade; as the Easterly Winds we had before shew'd that we were too far off the Land to have the Benefit of the Coasting South-Trade: and the faintness of both these Winds, and their often shifting from the S. S. W. to the S. E. with Squalls, Rain and small Gales, were a Confirmation of our being between the Verge of the S. Coasting-Trade, and that of the True Trade; which is here, regularly, S. E.

The third of *May* being in Lat. 20 deg. 00 min. and Merid. distance West from Cape *Salvadore* 234 Miles, the Variation was 7 deg. 00 min. We saw no Fowl but Shear-waters, as our Sea-men call them, being a small black Fowl that sweep the Water as they fly, and are much in the Seas that lie without either of the *Tropicks*: they are not eaten. We caught 3 small Sharks, each 6 Foot 4 Inches long; and they were very good Food for us. The next day we caught 3 more Sharks of the same size, and we eat them also, esteeming them as good Fish boil'd and prest, and then stew'd with Vinegar and Pepper.

We

An. 1699. We had nothing of Remark from the 3d of *May* to the 10th, only now and then seeing a small Whale spouting up the Water. We had the Wind Easterly, and we ran with it to the Southward, running in this time from the Lat. of 20 deg. 00 m. to 29 deg. 5 min. S. and having, then, 7 d. 3 m. E. Long. from C. *Salvadore*; the Variation increasing upon us, at present, notwithstanding we went East. We had all along a great difference between the Morning and Evening Amplitudes; usually a degree or two, and sometimes more. We were now in the True Trade, and therefore made good way to the Southward, to get without the Verge of the General Trade-Wind into a Westerly Wind's way, that might carry us towards the Cape of *Good Hope*. By the 12th of *May*, being in Lat. 31 deg. 10 min. we began to meet with Westerly Winds; which freshned on us, and did not leave us till a little before we made the Cape. Sometimes it blew so hard that it put us under a fore-course; especially in the Night: but in the day-time we had commonly our Main Top-sail rift. We met with nothing of moment; only we past by a dead Whale, and saw millions (as I may say) of Sea-Fowls about the Carkass (and as far round about it as we could see) some Feeding, and the rest flying about, or sitting

ting on the Water, waiting to take their Turns. We first discovered the Whale by the Fowls; for indeed I did never see so many Fowls at once in my Life before; their Numbers being inconceivably great: They were of divers sorts, in Bigness, Shape and Colour. Some were almost as big as Geese, of a grey Colour, with White Breasts, and with such Bills, Wings, and Tails. Some were *Pintado* Birds, as big as Ducks, and speckled Black and White. Some were Shear-waters; some Petrels; and there were several sorts of large Fowls. We saw of these Birds, especially the *Pintado*-birds, all the Sea over from about 200 Leagues distant from the Coast of *Brazil*, to within much the same distance of *New Holland*. The *Pintado* is a Southern Bird, and of that Temperate Zone; for I never saw of them much to the Norward of 30 deg. S. The *Pintado*-bird is as big as a Duck; but appears, as it flies, about the bigness of a tame Pigeon, having a short Tail, but the Wings very long, as most Sea-Fowls have; especially such as these that fly far from the Shore, and seldom come nigh it: for their Resting is sitting afloat upon the Water; but they lay, I suppose, ashore. There are three sorts of these Birds, all of the same make and bigness, and are only different in Colour. The first is black all over:

The

An. 1699. The second sort are grey, with white Bellies and Breasts. The third sort, which is the true *Pintado*, or Painted-bird, is curiously spotted white and black. Their Heads, and the tips of their Wings and Tails, are black for about an Inch; and their Wings are also edg'd quite round with such a small black List: only within the black on the tip of their Wings there is a white Spot seeming as they fly (for then their Spots are best seen) as big as a Half-crown. All this is on the outside of the Tails and Wings; and as there is a white Spot in the black Tip of the Wings, so there is in the middle of the Wings which is white, a black Spot; but this, towards the Back of the Bird, turns gradually to a dark grey. The Back its self, from the Head to the Tip of the Tail, and the Edge of the Wings next to the Back, are all over-spotted with fine small, round, white and black Spots, as big as a Silver Two-pence, and as close as they can stick one by another: The Belly, Thighs, Sides, and inner part of the Wings are of a light Grey. These Birds, of all these sorts, fly many together, never high, but almost sweeping the Water. We shot one awhile after on the Water in a Calm, and a Water-Spaniel we had with us brought it in: I have given a Picture of it [See *Birds*. Fig. 1.] but it was so damaged, that the Picture doth

Place this P. 96.

This very much resembles the Guarauna, described, and figured by Piso.

The Pintado Bird.
P. 96.

doth not ſhew it to advantage; and its Spots are beſt ſeen when the Feathers are ſpread as it flies.

The Petrel is a Bird not much unlike a Swallow, but ſmaller, and with a ſhorter Tail. 'Tis all over black, except a white Spot on the Rump. They fly ſweeping like Swallows, and very near the Water. They are not ſo often ſeen in fair Weather; being Foul-weather Birds, as our Seamen call them, and preſaging a Storm when they come about a Ship; who for that Reaſon don't love to ſee them. In a Storm they will hover cloſe under the Ship's Stern, in the Wake of the Ship (as 'tis call'd) or the ſmoothneſs which the Ship's paſſing has made on the Sea: and there as they fly (gently then) they pat the Water alternately with their Feet, as if they walkt upon it; tho' ſtill upon the Wing. And from hence the Seamen give them the name of *Petrels*, in alluſion to St. *Peter*'s walking upon the Lake of *Genneſareth*.

We alſo ſaw many Bunches of Sea-weeds in the Lat. of 39. 32. and by Judgment, near the Meridian of the Iſland *Triſtian d'Aconha*: and then we had about 2 d. 20 min. Eaſt Variation; which was now again decreaſing as we ran to the Eaſtward, till near the Meridian of *Aſcention*; where we found little or no Variation: But from thence, as we ran farther to the Eaſt, our Variation increaſed Weſterly.

An. 1699. Two days before I made the Cape of *G. Hope*, my Variation was 7 deg. 58 min. West. I was then in 43 deg. 27 min. East Longit. from C. *Salvador*, being in Lat. 35 deg. 30 min. this was the first of *June*. The second of *June* I saw a large black Fowl, with a whitish flat Bill, fly by us; and took great notice of it, because in the *East-India* Waggoner, or Pilot-book, there is mention made of large Fowls, as big as Ravens, with white flat Bills and black Feathers, that fly not above 30 Leagues from the *Cape*, and are lookt on as a Sign of ones being near it. My Reckoning made me then think my self above 90 Leagues from the *Cape*, according to the Longitude which the *Cape* hath in the common Sea-Charts: so that I was in some doubt, whether these were the right Fowls spoken of in the Waggoner; or whether those Fowls might not fly farther off Shore than is there mentioned; or whether, as it prov'd, I might not be nearer the *Cape* than I reckoned my self to be: for I found, soon after, that I was not then above 25 or 30 Leagues at most from the Cape. Whether the fault were in the Charts laying down the *Cape* too much to the East from *Brazil*, or were rather in our Reckoning, I could not tell: but our Reckonings are liable to such Uncertainties from Steerage, Log, Currents, Half

Minute-

Minute-Glasses, and sometimes want of Care, as in so long a Run cause often a difference of many Leagues in the whole Account. An. 1699.

Most of my Men that kept Journals imputed it to the Half-Minute-glasses; and indeed we had not a good Glass in the Ship beside the Half-watch or Two Hour-Glasses. As for our Half-Minute-Glasses we tried them all at several times, and we found those that we had us'd from *Brazil* as much too short, as others we had us'd before were too long: which might well make great Errors in those several Reckonings. A Ship ought therefore to have its Glasses very exact: and besides, an extraordinary Care ought to be used in heaving the Log, for fear of giving too much Stray-Line in a moderate Gale; and also to stop quickly in a brisk Gale; for when a Ship runs 8, 9, or 10 Knots, half a Knot or a Knot is soon run out, and not heeded: But to prevent danger, when a Man thinks himself near Land, the best way is to look out betimes, and lie by in the Night: for a Commander may err easily himself; beside the Errors of those under him, tho' never so carefully eyed.

Another thing that stumbled me here was the *Variation*, which, at this time, by the last Amplitude I had I found to be but 7 deg. 58 min. W. whereas the Variation

An. 1699. at the *Cape* (from which I found my ſelf not 30 Leagues diſtant) was then computed, and truly, about 11 Deg. or more: And yet a while after this, when I was got 10 Leagues to the Eaſtward of the *Cape*, I found the Variation but 10 Deg. 40 Min. W. whereas it ſhould have been rather more than at the *Cape*. Theſe Things, I confeſs, did puzzle me: neither was I fully ſatisfied as to the Exactneſs of the taking the Variation at Sea: For in a great Sea, which we often meet with, the Compaſs will traverſe with the motion of the Ship; beſides the Ship may and will deviate ſomewhat in Steering, even by the beſt Helmſmen: And then when you come to take an *Azimuth*, there is often ſome difference between him that looks at the Compaſs, and the Man that takes the Altitude heighth of the Sun; and a ſmall Error in each, if the Error of both ſhould be one way, will make it wide of any great Exactneſs. But what was moſt ſhocking to me, I found that the Variation did not always increaſe or decreaſe in proportion to the Degrees of Longitude Eaſt or Weſt; as I had a Notion they might do to a certain Number of Degrees of Variation Eaſt or Weſt, at ſuch or ſuch particular Meridians. But finding in this Voyage that the Difference of Variation did not bear a regular proportion to the difference of Longitude, I

was

was much pleas'd to see it thus Observ'd in a Scheme shewn me after my Return home, wherein are represented the several Variations in the *Atlantick* Sea, on both sides the Equator; and there, the Line of no Variation in that Sea is not a Meridian Line, but goes very oblique, as do those also which shew the Increase of Variation on each side of it. In that Draught there is so large an Advance made as well towards the Accounting for those seemingly Irregular Increases and Decreases of Variation towards the S. E. Coast of *America*, as towards the fixing a general Scheme or System of the Variation every where, which would be of such great Use in Navigation, that I cannot but hope that the Ingenious Author, Capt. *Halley*, who to his profound Skill in all Theories of these kinds, hath added and is adding continually Personal Experiments, will e'er long oblige the World with a fuller Discovery of the Course of the Variation, which hath hitherto been a Secret. For my part I profess my self unqualified for offering at any thing of a General Scheme; but since Matter of Fact, and whatever increases the History of the Variation, may be of use towards the setling or confirming the Theory of it, I shall here once for all insert a *Table* of all the *Variations* I observ'd beyond the *Equator* in this Voyage, both

An. 1699.

An. 1699. in going out, and returning back; and what Errors there may be in it, I shall leave to be Corrected by the Observations of Others.

A Table of Variations.

1699.		D. M. S. Lat.		D. M. Longit.		D. M. Variat.	
Mar.	14	6	15	1	47 a	3	27 E
	21	12	45	12	9	3	27
Apr.	25	14	49	00	10 b	7	0
	28	18	13	00	31	6	38
	30	19	00	2	20	6	30
May	2	19	22	3	51	8	15
	3	20	1	3	40	7	0
	5	22	47	3	48	9	40
	6	24	23	3	53	7	36
	7	25	44	3	53	10	15
	8	26	47	4	35	7	14
	9	28	9	5	50	9	45
	10	29	5	7	3	11	41
	11	29	23	7	38	12	47
	17	34	58	18	43	5	40
	18	34	54	19	06	6	19
	19	35	48	19	45	5	6
	23	39	42	27	1	2	55
	25	39	11	31	35	2	0
June	1	35	30	43	27	7	58 W

a W. from *St. Jago.*
b E. from C. *Salvador* in *Brazil.*

June

1699.		D. S. Lat.	M.	D. Longit.	M.	D. Variat.	M.
June	5	35	8	00	23 c	10	40 W
	6	36	7	3	6	11	10
	8	36	17	10	3	15	00
	9	35	59	12	0	19	38
	12	35	20	20	18	21	35
	14	35	5	26	13	23	50
	15	34	51	29	24	25	56
	17	34	27	36	8	24	54
	19	34	17	39	24	25	29
	20	34	15	42	25	24	22
	22	33	34	45	41	22	15
	25	35	8	45	28	24	30
	28	36	40	49	33	22	50
	29	36	40	53	12	22	44
	30	36	15	56	22	21	40
July	1	35	35	58	44	19	45
	4	33	32	66	22	16	40
	6	31	30	68	34	12	20
	7	31	45	69	00	12	2
	10	32	39	70	21	13	36
	11	33	4	72	00	12	29
	13	31	17	74	43	10	0
	15	29	20	75	25	10	28
	18	28	16	78	29	9	51
	23	26	43	84	19	9	11
	24	26	28	85	20	8	9
	25	26	14	85	52	8	40
	26	25	36	86	21	8	20

c E. from C. G. *Hope*.

1699.		D. M. S. Lat.		D. M. Longit.		D. M. Variat.		
July	27	26	43	86	16	7	0	W
	29	27	38	87	25	8	20	
	31	26	54	88	1	9	0	
Aug.	5	25	30	86	3	7	24	
	15	24	41	86	2 *d*	6	6	
	17	23	2	00	22	7	6	
	20	19	37	3	00	7	00	
	24	19	52	4	41	7	7	
	25	19	45	5	10	6	40	
	27	19	24	6	11	5	18	
	28	18	38	6	57	6	12	
Sept.	6	17	16	9	18	4	3	
	7	16	9	8	57	2	7	
	8	15	37	9	34	2	20	
	10	13	55	10	55	1	47	
	11	13	12	11	42	1	47	
Dec.	29	5	1	6	34 *e*	1[illegible]	2	E.
1700. *Jan.*	3	1	32	6	53	4	8	
Feb.	13	0	9	2	48 *f*	4	0	
	16	0	12	7	31	6	26	
	21	0	12	15	23	8	45	
	23	0	43	18	00	8	45	
	27	2	43	19	41	9	50	
Mar.	10	5	10	00	5 *g*	1	0	
	13	5	35	00	44 *h*	9	0	
	30	5	15	6	4	8	25	W.
Apr.	6	3	32	8	25	7	16	

d E. from *Sharks-Bay* in N. *Holland.*
e E. from *Babao*-Bay in J. *Timor.*
f E. from C. *Maba* in N. *Guinea.*
g E. from C. *St. George* on I. N. *Britannia.*
h W. from *ditto.*

April

An. 1699.

1700.		D.	M.	D.	M.		D.	M.	
		S. Lat.		Longit.			Variat.		
April	22	1	32	00	37	*i*	3	00	W
May	1	3	00			*k*	2	15	E
	24	9	59	00	25	*l*	0	15	W
	27	14	33	3	30		1	25	
June	2	19	44	8	7		5	38	
	3	19	51	9	58		6	10	
	4	19	46	11	6		6	20	
	5	20	00	12	22		4	58	
	6	20	00	14	17		7	20	
	9	19	59	16	01		6	32	
	11	9	57	17	42		8	1	
	12	19	48	19	0		6	0	
Nov.	7	21	26			*m*	9	0	
	14	27	1	35	35		16	50	
	15	27	10	36	34		18	57	
	16	27	11	37	54		17	24	
	19	28	14	41	40		19	39	
	21	29	24	44	47		20	50	
	23	29	42	47	34		21	38	
	24	30	16	49	26		26	00	
	25	30	40	51	24		22	38	
	27	31	51	55	5		22	40	
	29	32	55	56	28		27	10	
	30	31	55	57	25		27	10	
Dec.	1	31	57	58	17		24	30	
	2	31	57	59	33		27	57	
	4	32	3	61	45		24	50	

i W. from C. *Maba.*
k At Anchor off I. *Céram.*
l W. from *Babao*-Bay.
m W. from *Princes* Isle by *Java*-Head.

An. 1699.

1700.		D. M. S. Lat.		D. M. Longit.		D. M. Variat.	
		D.	M.	D.	M.	D.	M.
Dec.	6	32	15	66	00	23	30 W
	7	37	28	68	36	24	48
	8	33	49	64	38	21	53
	9	32	49	70	09	24	00
	11	32	50	71	45	21	15
	13	31	55	72	32	20	16
	14	31	35	73	39	20	00
	15	32	21	75	22	20	00
	17	33	5	79	39	18	42
	18	33	0	80	39	17	15
	21	34	39	82	46	16	41
	22	34	36	83	19	14	36
	23	34	21	83	42	14	00
	25	34	38	84	21	14	00
1701. *Jan.*	15	31	25	2	32 *	10	20
	16	30	5	4	42	9	36
	17	28	46	6	8	8	25
	18	27	26	7	32	7	40
	19	26	11	9	9	7	30
	20	25	00	10	49	7	9
	21	23	42	12	34	6	55
	22	22	51	14	10	5	56
	23	21	48	15	17	5	32
	24	21	24	15	51	4	56
	36	19	57	16	48	4	20
	27	19	10	17	22	3	24
	28	18	13	18	23	4	00
	29	17	22	19	29	2	00

* W. from the *Table*-Land at C. G. *Hope*.

Feb.

1701.		S. Lat. D. M.		Longit. D. M.		Variat. D. M.
Feb.	16	12 52		3 8 o		1 50 W
	17	11 55		4 42		1 10
	18	11 17		5 30		0 20
	19	10 22		6 32		1 10
	21	We made the I. *Ascention.*				

o W. from *Sanct Helena.*

But

1699. But to return from this Digression: Having fair Weather, and the Winds hanging Southerly, I jog'd on to the Eastward, to make the *Cape*. On the third of *June* we saw a Sail to Leeward of us, shewing *English* Colours. I bore away to speak with her, and found her to be the *Antelope* of *London*, commanded by Captain *Hammond*, and bound for the Bay of *Bengal* in the Service of the *New-East-India* Company. There were many Passengers aboard, going to settle there under Sir *Edward Littleton*, who was going Chief thither: I went aboard, and was known by Sir *Edward*, and Mr. *Hedges*, and kindly received and treated by them and the Commander; who had been afraid of us before, tho' I had sent one of my Officers aboard. They had been in at the *Cape*, and came from thence the Day before, having stockt themselves with Refreshments. They told me that they were by Reckoning, 60 Miles to the West of the *Cape*. While I was aboard them, a fine small Westerly Wind sprang up; therefore I shortned my stay with them, because I did not design to go in to the *Cape*. When I took leave I was presented with half a Mutton, 12 Cabbages, 12 Pumkins, 6 Pound of Butter, 6 Couple of Stockfish, and a quantity of Parsnips; sending them some Oatmeal, which they wanted.

From

From my first setting out from *England*, An. 1699.
I did not design to touch at the *Cape*; and that was one Reason why I touch'd at *Brazil*, that there I might refresh my Men, and prepare them for a long Run to *New Holland*. We had not yet seen the Land; but about 2 in the Afternoon we saw the *Cape*-Land bearing East, at above 16 Leagues distance: And Captain *Hammond* being also bound to double the *Cape*, we jog'd on together this Afternoon and the next Day, and had several fair Sights of it; which may be seen [*Table* III. No. 6. 7. 8.]

To proceed, having still a Westerly Wind, I jog'd on in company with the *Antelope*, till Sunday *June* the 4th at 4 in the Afternoon, when we parted; they steering away for the *East-Indies*, and I keeping an E. S. E. Course, the better to make my way for *New Holland*: For tho' *New Holland* lies North-Easterly from the *Cape*, yet all Ships bound towards that Coast, or the Streights of *Sundy*, ought to keep for a while in the same Parallel, or in a Lat. between 35 and 40. at least a little to the S. of the East, that they may continue in a variable Winds way; and not venture too soon to stand so far to the North, as to be within the verge of the Trade-Wind, which will put them by their Easterly Course. The Wind increased upon us; but

we

An. 1699. we had yet ſight of the *Antelope*, and of the Land too, till *Tueſday* the 6th of *June*: And then we ſaw alſo by us an inumerable Company of Fowls of divers ſorts; ſo that we lookt about to ſee if there were not another dead Whale, but ſaw none.

The Night before, the Sun ſet in a black Cloud, which appeared juſt like Land; and the Clouds above it were gilded of a dark red Colour. And on the *Tueſday*, as the Sun drew near the Horizon, the Clouds were gilded very prettily to the Eye, tho' at the ſame time my Mind dreaded the Conſequences of it. When the Sun was now not above 2 deg. high, it entered into a dark ſmoaky-coloured Cloud that lay parallel with the Horizon, from whence preſently ſeem'd to iſſue many dusky blackiſh Beams. The Sky was at this time covered with ſmall hard Clouds (as we call ſuch as lie ſcattering about, not likely to Rain) very thick one by another; and ſuch of them as lay next to the Bank of Clouds at the Horizon, were of a pure Gold colour to 3 or 4 deg. high above the Bank: From theſe to about 10 deg. high they were redder, and very bright; above them they were of a darker Colour ſtill, to about 60 or 70 deg. high; where the Clouds began to be of their common Colour. I took the more particular Notice of all this, becauſe I have generally obſerved ſuch colour'd Clouds to appear

pear before an approaching Storm: And this being Winter here, and the time for bad Weather, I expected and provided for a violent Blast of Wind, by riffing our Topsails, and giving a strict charge to my Officers to hand them or take them in, if the Wind should grow stronger. The Wind was now at W. N. W. a very brisk Gale. About 12 a Clock at Night we had a pale whitish Glare in the N. W. which was another Sign, and intimated the Storm to be near at hand; and the Wind increasing upon it, we presently handed our Topsails, furled the Mainsail, and went away only with our Foresail. Before 2 in the Morning it came on very fierce, and we kept right before Wind and Sea, the Wind still increasing: But the Ship was very governable, and Steer'd incomparably well. At 8 in the Morning we settled our Fore-Yard, lowering it 4 or 5 Foot, and we ran very swiftly; especially when the Squals of Rain or Hail, from a black Cloud, came over head, for then it blew excessive hard. These, tho' they did not last long, yet came very thick and fast one after another. The Sea also ran very high: But we running so violently before Wind and Sea, we Shipt little or no Water; tho' a little washt into our upper Deck-Ports; and with it a Scuttle or Cuttle-Fish was cast upon the Carriage of a Gun.

An. 1699.

The

An. 1699. The Wind blew extraordinary hard all *Wednesday*, the 7th of *June*, but abated of its fierceness before Night: Yet it continued a brisk Gale till about the 16th, and still a moderate one till the 19th Day; by which time we had run about 600 Leagues: For the most part of which time the Wind was in some point of the West, *viz.* from the W. N. W. to the S. by W. It blew hardest when at W. or between the W. and S. W. but after it veered more Southerly the foul Weather broke up: This I observed at other times also in these Seas, that when the Storms at West veered to the Southward they grew less; and that when the Wind came to the E. of the S. we had still smaller Gales, Calms, and fair Weather. As for the Westerly Winds on that side the *Cape*, we like them never the worse for being violent, for they drive us the faster to the Eastward; and are therefore the only Winds coveted by those who sail towards such parts of the *East-Indies*, as lie South of the Equator; as *Timor*, *Java*, and *Sumatra*; and by the Ships bound for *China*, or any other that are to pass through the Streights of *Sundy*. Those Ships having once past the *Cape*, keep commonly pretty far Southerly, on purpose to meet with these West Winds, which in the Winter Season of these Climates they soon meet with; for then the Winds are generally Westerly at the *Cape*, and especially to the Southward of it: But in their Summer Months they get to the Southward of 40 deg. usually e'er they meet with the Westerly Winds. I was not at this time in a higher Lat. than 36 deg. 40 min. and oftentimes was more Northerly, altering my Latitude often as Winds and Weather requir'd; for in such long Runs 'tis best to shape ones Course according to the Winds. And if in Steering to the East, we should be obliged to bear a little to the N. or S. of it, 'tis no great matter; for 'tis but Sailing 2 or 3 Points from the Wind, when 'tis either Northerly or Southerly; and this not only easeth the Ship from straining, but shortens the way more than if a Ship was kept close on a Wind, as some Men are fond of doing.

An. 1699.

The 19th of *June* we were in Lat. 34 deg. 17 min. S. and Long. from the *Cape* 39 deg. 24 min. E. and had small Gales and Calms. The Winds were at N. E. by E. and continued in some part of the E. till the 27th Day. When it having been some time at N. N. E. it came about at N. and then to the W. of the N. and continued in the West-board (between the N. N. W. and S. S. W.) till the 4th of *July*; in which time we ran 782 Miles; then the Winds came about again to the East, we reckoning our selves to be in a Meridian 1100 L. East of that of the *Cape*; and having fair Weather sounded, but had no Ground.

An. 1699. We met with little of Remark in this Voyage, besides being accompanied with Fowles all the way, especially Pintado-Birds, and seeing now and then a Whale: But as we drew nigher the Coast of *New-Holland*, we saw frequently 3 or 4 Whales together. When we were about 90 Leagues from the Land we began to see Sea-weeds, all of one sort; and as we drew nigher the Shore we saw them more frequently. At about 30 Leagues distance we began to see some Scutle-bones floating on the Water; and drawing still nigher the Land we saw greater quantities of them.

July the 25th being in Lat. 26 deg. 14 min. S. and Longitude E. from the C. of *G. Hope* 85 deg. 52 min. we saw a large Gar-fish leap 4 times by us, which seemed to be as big as a Porpose. It was now very fair Weather, and the Sea was full of a sort of very small Grass or Moss, which as it floated in the Water seemed to have been some Spawn of Fish; and there was among it some small Fry. The next Day the Sea was full of small round things like Pearl, some as big as white Peas; they were very Clear and Transparent, and upon crushing any of them a drop of Water would come forth: The Skin that contained the Water was so thin that it was but just deseernable. Some Weeds swam by us, so that we did not doubt but we should quickly see Land.

On

On the 27th alſo, ſome Weeds ſwam by us, and the Birds that had flown along with us all the way almoſt from *Brazil*, now left us, except only 2 or 3 Shear-waters. On the 28th we ſaw many Weeds ſwim by us, and ſome Whales, blowing. On the 29th we had dark cloudy Weather, with much Thunder, Lightning, and violent Rains in the Morning: But in the Evening it grew fair. We ſaw this Day a Scutle-bone ſwim by us, and ſome of our young Men a Seal, as it ſhould ſeem by their Deſcription of its Head. I ſaw alſo ſome Boneta's, and ſome Skipjacks, a Fiſh about 8 Inches long, broad and ſizable; not much unlike a Roach; which our Seamen call ſo from their leaping about.

An. 1699.

The 30th of *July*, being ſtill nearer the Land, we ſaw abundance of Scutle-bones and Sea-weed, more Tokens that we were not far from it; and ſaw alſo a ſort of Fowls the like of which we had not ſeen in the whole Voyage, all the other Fowls having now left us. Theſe were as big as Lapwings; of a grey Colour, black about their Eyes, with red ſharp Bills, long Wings, their Tails long and forked like Swallows; and they flew flapping their Wings like Lapwings. In the Afternoon we met with a Ripling like a Tide or Current, or the Water of ſome Shole or Overfal; but were paſt it before we could ſound.

An. 1699. The Birds last mention'd and this were further Signs of Land. In the Evening we had fair Weather, and a small Gale at West. At 8 a Clock we sounded again; but had no Ground.

We kept on still to the Eastward, with an easy Sail, looking out sharp: for by the many Signs we had, I did expect that we were near the Land. At 12 a Clock in the Night I sounded, and had 45 Fathom, course Sand and small white Shells. I presently clapt on a Wind and stood to the South, with the Wind at W. because I thought we were to the South of a Shoal call'd the *Abrohles* (an Appellative Name for Shoals, as it seems to me) which in a Draught I had of that Coast is lay'd down in 27 deg. 28 min. Lat. stretching about 7 Leagues into the Sea. I was the Day before in 27 deg. 38 min. by Reckoning. And afterwards steering E. by S. purposely to avoid it, I thought I must have been to the South of it: but sounding again, at One a Clock in the Morning, *Aug*. the first, we had but 25 Fathom, Coral-Rocks; and so found the Shoal was to the South of us. We presently tackt again, and stood to the North, and then soon deepned our Water; for at two in the Morning we had 26 Fathom Coral still: At three we had 28 Coral-ground: At 4 we had 30 Fathom, course Sand, with some Coral: At

5 we

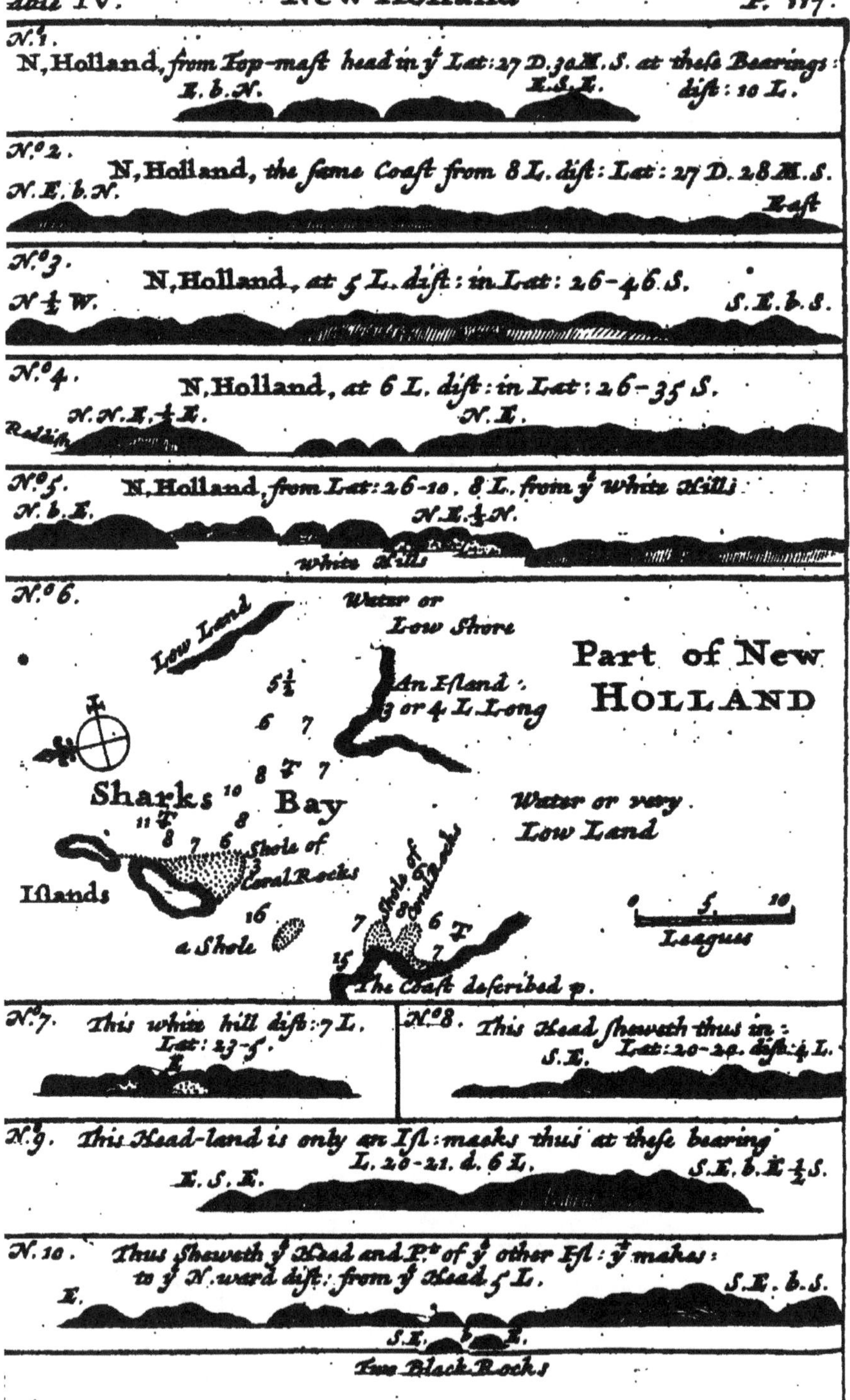
N.º 1.
N, Holland, from Top-mast head in ye Lat: 27 D. 30 M. S. at these Bearings:
E. b. N.
E. S. E.
dist: 10 L.
N.º 2.
N, Holland, the same Coast from 8 L. dist: Lat: 27 D. 28 M. S.
N. E. b. N.
East
N.º 3.
N, Holland, at 5 L. dist: in Lat: 26-46 S.
N ½ W.
S. E. b. S.
N.º 4.
N, Holland, at 6 L. dist: in Lat: 26-35 S.
N. N. E. ½ E.
N. E.
N.º 5.
N, Holland, from Lat: 26-10. 8 L. from ye White Hills
N. b. E.
N. E. ½ N.
White Hills
N.º 6.
Low Land
Water or Low Shore
Part of New HOLLAND
An Island 3 or 4 L. Long
Sharks Bay
Shole of Coral Rocks
Islands
a Shole
Water or very Low Land
Leagues
The Coast described
N.º 7. This white hill dist: 7 L. Lat: 23-5.
N.º 8. This Head sheweth thus in: Lat: 20-20. dist: 4 L.
S. E.
N.º 9. This Head-land is only an Isl: makes thus at these bearing
L. 20-21. d. 6 L.
E. S. E.
S. E. b. E ½ S.
N. 10. Thus Sheweth ye Head and Pt of ye other Isl: ye makes: to ye N. ward dist: from ye Head 5 L.
E.
S. E. b. S.
S. E.
E.
Two Black Rocks

5 we had 45 Fathom, course Sand and Shells; being now off the Shole, as appear'd by the Sand and Shells, and by having left the Coral. By all this I knew we had fall'n in to the North of the Shole, and that it was laid down wrong in my Sea-Chart: for I found it lie in about 27 deg. Lat. and by our Run in the next day, I found that the Outward-edge of it, which I sounded on; lies 16 Leagues off Shore. When it was day we steered in E. N. E. with a fine brisk Gale; but did not see the Land till 9 in the Morning, when we saw it from our Topmast-head, and were distant from it about 10 Leagues; having then 40 Fathom-water, and clean Sand. About 3 Hours after we saw it on our Quarter-Deck, being by Judgment about 6 Leagues off: and we had then 40 Fathom, clean Sand. As we ran in, this day and the next, we took several Sights of it, at different Bearings and Distances; from which it appear'd as you see in [*Table* IV. N°. 1, 2, 3, 4, 5.] And here I would Note once for all, That the Latitudes mark'd in the Draughts, or Sights here given, are not the Latitude of the Land, but of the Ship when the Sight was taken. This Morning, *August* the first, as we were standing in we saw several large Sea-fowls, like our Gannets on the Coast of *England*, flying three or four together;

An. 1699.

An. 1699. gether; and a sort of white Sea-Mews, but black about the Eyes, and with forked Tails. We strove to run in near the Shore to seek for a Harbour to refresh us after our tedious Voyage; having made one continued stretch from *Brazil* hither of about 114 Deg.; designing from hence also to begin the Discovery I had a mind to make on *N. Holland* and *N. Guinea.* The Land was low, and appear'd even, and as we drew nearer to it, it made (as you see in *Table* IV. N°. 3, 4, 5.) with some red and some white Clifts; these last in Lat. 26. 10 S. where you will find 54 Fathom, within four Miles of the Shore.

About the Lat. of 26 deg. S. we saw an Opening, and ran in, hoping to find a Harbour there: but when we came to its Mouth, which was about two Leagues wide, we saw Rocks and foul Ground within, and therefore stood out again: There we had 20 Fathom-water within two mile of the Shore. The Land every where appear'd pretty low, flat and even; but with steep Cliffs to the Sea; and when we came near it there were no Trees, Shrubs or Grass to be seen. The Soundings in the Lat. of 26 deg. S. from about 8 or 9 Leagues off till you come within a League of the Shore, are generally about 40 Fathom; differing but little, seldom above three or four Fathom. But the

Lead

Lead brings up very different sorts of Sand, some course, some fine; and of several Colours, as Yellow, White, Grey, Brown, Bluish and Reddish. *An.1699.*

When I saw there was no Harbour here, nor good Anchoring, I stood off to Sea again, in the Evening of the second of *August*, fearing a Storm on a Lee-shore, in a place where there was no shelter, and desiring at least to have Sea-Room: For the Clouds began to grow thick in the Western-board, and the Wind was already there, and began to blow fresh almost upon the Shore; which at this Place lies along N. N. W. and S. S. E. By Nine a Clock at Night we had got a pretty good Offin; but the Wind still increasing, I took in my Main Top-sail, being able to carry no more Sail than two Courses and the Mizen. At two in the Morning, *Aug.* 3. it blew very hard, and the Sea was much raised; so that I furled all my Sails but my Main-sail. Tho' the Wind blew so hard, we had yet pretty clear Weather till Noon: But then the whole Sky was blackned with thick Clouds, and we had some Rain, which would last a quarter of an hour at a time, and then it would blow very fierce while the Squals of Rain were over our Heads; but as soon as they were gone the Wind was by much abated, the stress of the Storm being over. We sound-

I 4 ed

An. 1699. ed ſeveral times, but had no Ground till 8 a Clock *Aug.* the 4th. in the Evening; and then had 60 Fathom-water, Coral-ground. At Ten we had 56 Fathom fine Sand. At Twelve we had 55 Fathom, fine Sand, of a pale, bluiſh Colour. It was now pretty moderate Weather; yet I made no Sail till Morning: but then, the Wind veering about to the S. W. I made Sail and ſtood to the North: And at 11 a Clock the next day, *Aug.* 5. we ſaw Land again, at about 10 Leagues diſtance. This Noon we were in Lat. 25 deg. 30 min. and in the Afternoon our Cook died, an Old Man, who had been ſick a great while, being infirm before we came out of *England*.

The 6th of *Auguſt* in the Morning we ſaw an Opening in the Land, and we ran in to it and anchored in ſeven and a half Fathom-water, 2 miles from the Shore, clean Sand. It was ſomewhat difficult getting in here, by reaſon of many Shoals we met with: But I ſent my Boat ſounding before me. The Mouth of this Sound, which I call'd *Shark*'s *Bay*, lies in about 25 deg. S. Lat. and our Reckoning made its Longitude from the C. of *Good Hope* to be about 87 Degrees; which is leſs by 195 Leagues than is uſually laid down in our common Draughts, if our Reckoning was right, and our Glaſſes did not deceive us.

us. As ſoon as I came to anchor in this Bay (of which I have given a Plan, Table IV. N°. 6.) I ſent my Boat aſhore to ſeek for freſh Water: But in the Evening my Men returned, having found none. The next morning I went aſhore my ſelf, carrying Pick-axes and Shovels with me, to dig for Water; and Axes to cut Wood. We tried in ſeveral places for Water, but finding none after ſeveral Trials, nor in ſeveral miles compaſs, we left any farther ſearch for it, and ſpending the reſt of the day in cutting Wood, we went aboard at Night.

An. 1699.

The Land is of an indifferent heighth, ſo that it may be ſeen 9 or 10 Leagues off. It appears at a diſtance very even; but as you come nigher you find there are many gentle Riſings, tho' none ſteep nor high. 'Tis all a ſteep Shore againſt the open Sea: but in this Bay or Sound we were now in, the Land is low by the Sea-ſide, riſing gradually in within the Land. The Mould is Sand by the Sea-ſide, producing a large ſort of Sampier, which bears a white Flower. Farther in, the Mould is reddiſh, a ſort of Sand producing ſome Graſs, Plants, and Shrubs. The Graſs grows in great Tufts, as big as a Buſhel, here and there a Tuft: being intermix'd with much Heath, much of the kind we have growing on our Commons in *England*.

An. 1699. *land.* Of Trees or Shrubs here are divers ſorts; but none above ten Foot high: Their Bodies about 3 Foot about, and 5 or 6 Foot high before you come to the Branches, which are buſhy and compos'd of ſmall Twigs there ſpreading abroad, tho' thick ſet, and full of Leaves; which were moſtly long and narrow. The Colour of the Leaves was on one ſide Whitiſh, and on the other Green: and the Bark of the Trees was generally of the ſame Colour with the Leaves; of a pale Green. Some of theſe Trees were ſweet-ſcented, and reddiſh within the Bark, like Saſſafras, but redder. Moſt of the Trees and Shrubs had at this time either Bloſſoms or Berries on them. The Bloſſoms of the different ſort of Trees were of ſeveral Colours, as Red, White, Yellow, *&c.* but moſtly Blue: and theſe generally ſmelt very ſweet and fragrant, as did ſome alſo of the reſt. There were alſo beſide ſome Plants, Herbs, and tall Flowers, ſome very ſmall Flowers, growing on the Ground, that were ſweet and beautiful, and for the moſt part unlike any I had ſeen elſewhere.

There were but few Land-Fowls: we ſaw none but Eagles, of the larger ſorts of Birds; but 5 or 6 ſorts of ſmall Birds. The biggeſt ſort of theſe were not bigger than Larks; ſome no bigger than Wrens, all ſing-

Place this P. 123.

The head & greatest part of yͤ neck of this bird is red, & therein differs from the Avosetta of Italy.

A Noddy of N. Holland. P.123 & 14

A Comon Noddy. P. 14

The Bill & Leggs of this Bird are of a Bright Red.

ſinging with great variety of fine ſhrill Notes; and we ſaw ſome of their Neſts with young Ones in them. The Water-Fowls are Ducks, (which had young Ones now, this being the beginning of the Spring in theſe Parts;) Curlews, Galdens, Crab-catchers, Cormorants, Gulls, Pelicans; and ſome Water-Fowl, ſuch as I have not ſeen any where beſides. I have given the Pictures of 4 ſeveral Birds on this Coaſt. [See *Birds*: Fig. 2, 3, 4, 5.]

An. 1699.

The Land-Animals that we ſaw here were only a ſort of Raccoons, different from thoſe of the *West-Indies*, chiefly as to their Legs; for theſe have very ſhort fore Legs; but go Jumping upon them as the others do, and like them are very good Meat:) and a ſort of Guano's, of the ſame ſhape and ſize with other Guano's, deſcrib'd [Vol. I. p. 57.] but differing from them in three remarkable Particulars: For theſe had a larger and uglier Head; and had no Tail: And at the Rump, inſtead of the Tail there, they had a ſtump of a Tail, which appear'd like another Head; but not really ſuch, being without Mouth or Eyes: Yet this Creature ſeem'd by this means to have a Head at each end; and, which may be reckon'd a fourth difference, the Legs alſo ſeem'd all four of them to be Fore-legs, being all alike in ſhape and length, and ſeeming by the

An. 1699. the Joints and Bending to be made as if they were to go indifferently either Head or Tail foremost. They were speckled black and yellow like Toads, and had Scales or Knobs on their Backs like those of Crocodiles, plated on to the Skin, or stuck into it, as part of the Skin. They are very slow in motion; and when a Man comes nigh them they will stand still and hiss, not endeavouring to get away. Their Livers are also spotted black and yellow: and the Body when opened hath a very unsavory Smell. I did never see such ugly Creatures any where but here. The Guano's I have observ'd to be very good Meat: and I have often eaten of them with pleasure: But tho' I have eaten of Snakes, Crocodiles and Allegators, and many Creatures that look frightfully enough, and there are but few I should have been afraid to eat of if prest by Hunger, yet I think my Stomach would scarce have serv'd to venture upon these *N. Holland* Guano's, both the Looks and the Smell of them being so offensive.

The Sea-fish that we saw here (for here was no River, Land or Pond of Fresh Water to be seen) are chiefly Sharks. There are abundance of them in this particular Sound, that I therefore gave it the Name of *Shark*'s *Bay*. Here are also Skates, Thornbacks, and other Fish of the Ray-kind;

kind; (one sort especially like the Sea-Devil) and Garfish, Boneta's, *&c.* Of Shell-fish we got here Muscles, Periwinkles, Limpits, Oysters, both of the Pearl-kind and also Eating-Oysters, as well the common sort as long Oysters; beside Cockles, *&c.* The Shore was lined thick with many other sorts of very strange and beautiful Shells, for variety of Colour and Shape, most finely spotted with Red, Black, or Yellow, *&c.* such as I have not seen any where but at this place. I brought away a great many of them; but lost all, except a very few, and those not of the best.

There are also some green Turtle weighing about 200 ℔. Of these we caught 2 which the Water Ebbing had left behind a Ledge of Rock, which they could not creep over. These served all my Company 2 Days; and they were indifferent sweet Meat. Of the Sharks we caught a great many, which our Men eat very savourily. Among them we caught one which was 11 Foot long. The space between its 2 Eyes was 20 Inches, and 18 Inches from one Corner of his Mouth to the other. Its Maw was like a Leather Sack, very thick, and so tough that a sharp Knife could scarce cut it: In which we found the Head and Boans of a *Hippopotomus*; the hairy Lips of which were still sound and not putrified,

and

An. 1699. and the Jaw was also firm, out of which we pluckt a great many Teeth, 2 of them 8 Inches long, and as big as a Mans Thumb, small at one end, and a little crooked; the rest not above half so long. The Maw was full of Jelly which stank extreamly: However I saved for a while the Teeth and the Sharks Jaw: The Flesh of it was divided among my Men; and they took care that no waste should be made of it.

'Twas the 7th of *August* when we came into *Shark*'s Bay; in which we Anchor'd at three several Places, and stay'd at the first of them (on the W. side of the Bay) till the 11th. During which time we searched about, as I said, for fresh Water, digging Wells, but to no purpose. However, we cut good store of Fire-wood at this first Anchoring-place; and my Company were all here very well refreshed with Raccoons, Turtle, Shark and other Fish, and some Fowles; so that we were now all much brisker than when we came in hither. Yet still I was for standing farther into the Bay, partly because I had a Mind to increase my stock of fresh Water, which was began to be low; and partly for the sake of Discovering this part of the Coast. I was invited to go further, by seeing from this Anchoring-place all open before me; which therefore I designed to search before I left the Bay. So on the 11th about Noon, I steer'd

I steer'd farther in, with an easie Sail, because we had but shallow Water: We kept therefore good looking out for fear of Sholes; sometimes shortning, sometimes deepning the Water. About 2 in the Afternoon we saw the Land a Head that makes the S. of the Bay, and before Night we had again Sholdings from that Shore: And therefore shortned Sail and stood off and on all Night, under 2 Topsails, continually sounding, having never more then 10 Fathom, and seldom less than 7. The Water deepned and sholdned so very gently, that in heaving the Lead 5 or 6 times we should scarce have a Foot difference. When we came into 7 Fathom either way, we presently went about. From this S. part of the Bay, we could not see the Land from whence we came in the Afternoon: And this Land we found to be an Island of 3 or 4 Leagues long, as is seen in the Plain, [Table IV. No. 6.] but it appearing barren, I did not strive to go nearer it; and the rather because the Winds would not permit us to do it without much Trouble, and at the Openings the Water was generally Shole. I therefore made no farther attempts in this S. W. and S. part of the Bay, but steered away to the Eastward, to see if there was any Land that way, for as yet we had seen none there. On the 12th in the Morning we pass'd by the N. Point of that

An. 1699.

An. 1699. that Land, and were confirm'd in the Perſuaſion of its being an Iſland, by ſeeing an Opening to the Eaſt of it, as we had done on the W. Having fair Weather, a ſmall Gale and ſmooth Water, we ſtood further on in the Bay, to ſee what Land was on the E. of it. Our Soundings at firſt were 7 Fathom, which held ſo a great while, but at length it decreas'd to 6. Then we ſaw the Land right a-head, that in the Plan makes the E. of the Bay. We could not come near it with the Ship, having but Shole-water: and it being dangerous lying there, and the Land extraordinarily low, very unlikely to have freſh Water (though it had a few Trees on it, ſeemingly Mangroves) and much of it probably covered at High-water, I ſtood out again that Afternoon, deepning the Water, and before Night anchored in 8 Fathom, clean white Sand, about the middle of the Bay. The next day we got up our Anchor; and that Afternoon came to an Anchor once more near two Iſlands, and a Shole of Corral Rocks that face the Bay. Here I ſcrubb'd my Ship: and finding it very improbable I ſhould get any thing further here, I made the beſt of my way out to Sea again, ſounding all the way: but finding by the ſhallowneſs of the Water that there was no going out to Sea to the Eaſt of the two Iſlands that face the

Bay,

Bay, nor between them, I return'd to the West Entrance, going out by the same Way I came in at, only on the East instead of the West-side of the small Shole to be seen in the Plan: in which Channel we had 10, 12, and 13 Fathom-water, still deepning upon us till we were out at Sea. The day before we came out I sent a Boat ashore to the most Northerly of the Two Islands, which is the least of them, catching many small Fish in the mean while with Hook and Line. The Boat's Crew returning, told me, That the Isle produces nothing but a sort of green, short, hard, prickly Grass, affording neither Wood nor fresh Water; and that a Sea broak between the two Islands, a Sign that the Water was shallow. They saw a large Turtle, and many Skates and Thornbacks, but caught none.

An. 1699.

It was *August* the 14th when I sail'd out of this Bay or Sound, the Mouth of which lies, as I said, in 25 deg. 5 min. designing to coast along to the N. E. till I might commodiously put in at some other part of *N. Holland*. In passing out we saw three Water-Serpents swimming about in the Sea, of a yellow Colour, spotted with dark, brown Spots. They were each about four Foot long, and about the bigness of a Man's Wrist, and were the first I saw on this Coast, which abounds with

An. 1699. several sorts of them. We had the Winds at our first coming out at N. and the Land lying North-Easterly. We plied off and on, getting forward but little till the next day: When the Wind coming at S. S. W. and S. we began to Coast it along the Shore to the Northward, keeping at 6 or 7 Leagues off Shore; and sounding often, we had between 40 and 46 Fathom-water, brown Sand, with some white Shells. This 15th of *August* we were in Lat. 24 deg. 41 min. On the 16th Day at Noon we were in 23 deg. 22 min. The Wind coming at E. by N. we could not keep the Shore aboard, but were forced to go farther off, and lost sight of the Land. Then sounding we had no Ground with 80 Fathom-line; however the Wind shortly after came about again to the Southward, and then we jogg'd on again to the Northward, and saw many small Dolphins and Whales, and abundance of Scuttle-shells swimming on the Sea; and some Water-snakes every day. The 17th we saw the Land again, and took a Sight of it. [See Table IV. N°. 7.]

The 18th in the Afternoon, being 3 or 4 Leagues off Shore, I saw a Shole-point, stretching from the Land into the Sea, a League or more. The Sea broke high on it; by which I saw plainly there was a Shole there. I stood farther off, and coast-

ed

An. 1699.

ed along Shore, to about 7 or 8 Leagues distance: And at 12 a Clock at Night we sounded, and had but 20 Fathom, hard Sand. By this I found I was upon another Shole, and so presently steered off W. half an hour, and had then 40 Fathom. At One in the Morning of the 18th day we had 85 Fathom: By Two we could find no Ground; and then I ventur'd to steer along Shore again, due N. which is two Points wide of the Coast (that lies here N. N. E.) for fear of another Shole. I would not be too far off from the Land, being desirous to search into it where-ever I should find an Opening or any Convenience of searching about, for Water, *&c.* When we were off the Shole-point I mention'd where we had but 20 Fathom-water, we had in the Night abundance of Whales about the Ship, some a head, others a-stern, and some on each side blowing and making a very dismal Noise; but when we came out again into deeper Water they left us. Indeed the Noise that they made by blowing and dashing of the Sea with their Tails, making it all of a Breach and Fome, was very dreadful to us, like the breach of the Waves in very Shole-water, or among Rocks. The Shole these Whales were upon had depth of Water sufficient, no less than twenty Fathom, as I said; and it lies in Lat. 22

An. 1699. deg. 22 min. The Shore was generally bold all along: we had met with no Shole at Sea since the *Abrohlo*-shole, when we first fell on the *N. Holland* Coast in the Lat. of 28. till yesterday in the Afternoon, and this Night. This Morning also when we expected by the Draught we had with us to have been 11. Leagues off Shore, we were but 4: so that either our Draughts were faulty, which yet hitherto and afterwards we found true enough as to the lying of the Coast, or else here was a Tide unknown to us that deceived us; tho' we had found very little of any Tide on this Coast hitherto. As to our Winds in the Coasting thus far, we had been within the Verge of the General Trade (tho' interrupted by the Storm I mention'd) from the Lat. of 28, when we first fell in with the Coast: and by that time we were in the Lat. of 25. we had usually the regular Trade-wind (which is here S. S. E.) when we were at any distance from Shore: but we had often Sea and Land-Breezes, especially when near Shore, and when in *Sharks-bay*; and had a particular N. West Wind, or Storm, that set us in thither. On this 18th of *August* we coasted with a brisk Gale of the True Trade-wind at S. S. E. very fair and clear Weather; but haling off in the Evening to Sea, were next Morning out of sight of Land: and the

Land

Land now trending away N. Easterly, and we being to the Norward of it, and the Wind also shrinking from the S. S. E. to the E. S. E. (that is, from the True Trade-Wind to the Sea-Breeze, as the Land now lay) we could not get in with the Land again yet a-while, so as to see it, tho' we trim'd sharp and kept close on a Wind. We were this 19th day in Lat. 21 deg. 42 min. The 20th we were in Lat. 19 deg. 37 min. and kept close on a Wind to get sight of the Land again, but could not yet see it. We had very fair Weather; and tho' we were so far from the Land as to be out of sight of it, yet we had the Sea and Land-Breezes. In the Night we had the Land-breeze at S. S. E. a small gentle Gale; which in the Morning about Sun-rising would shift about gradually (and withal increasing in Strength) till about Noon we should have it at E. S. E. which is the true Sea-breeze here. Then it would blow a brisk Gale, so that we could scarce carry our Top-sails double rift: and it would continue thus till 3 in the Afternoon, when it would decrease again. The Weather was fair all the while, not a Cloud to be seen; but very hazy, especially nigh the Horizon. We sounded several times this 20th day, and at first had no Ground: but had afterwards from 52 to 45 Fathom, course

An. 1699.

*An.*1699. brown Sand, mixt with small, brown and white Stones, with Dints besides in the Tallow.

The 21st day also we had small Land-breezes in the Night, and Sea-breezes in the day: and as we saw some Sea-snakes every day, so this day we saw a great many, of two different sorts or shapes. One sort was yellow, and about the bigness of a Man's Wrist, about 4 Foot long, having a flat Tail about 4 Fingers broad. The other sort was much smaller and shorter, round and spotted black and yellow. This day we sounded several times, and had 45 Fathom, Sand. We did not make the Land till Noon, and then saw it first from our Topmast-head. It bore S.E. by E. about 9 Leagues distance; and it appeared like a Cape or Head of Land. The Sea-breeze this day was not so strong as the day before, and it veered out more; so that we had a fair Wind to run in with to the Shore, and at Sun-set anchored in 20 Fathom, clean Sand, about 5 Leagues from the bluff Point; which was not a Cape (as it appear'd at a great distance) but the Eastermost end of an Island, about 5 or 6 Leagues in length, and one in breadth. There were 3 or 4 Rocky Islands about a League from us between us and the bluff Point; and we saw many other Islands both to the East and West of it, as

far

far as we could see either way from our Topmast-head: And all within them to the S. there was nothing but Islands of a pretty heighth, that may be seen 8 or 9 Leagues off. By what we saw of them they must have been a Range of Islands of about 20 Leagues in length, stretching from E. N. E. to VV. S. VV. and for ought I know, as far as to those of *Sharks-Bay*; and to a considerable breadth also, (for we could see 9 or 10 Leagues in among them) towards the Continent or main Land of *N. Holland*, if there be any such thing hereabouts: and by the great Tides I met with awhile afterwards, more to the N. East, I had a strong suspicion that here might be a kind of *Archipelago* of Islands, and a Passage possibly to the S. of *N. Holland* and *N. Guinea* into the great *S. Sea* Eastward; which I had Thoughts also of attempting in my Return from *N. Guinea* (had Circumstances permitted) and told my Officers so: but I would not attempt it at this time, because we wanted VVater, and could not depend upon finding it there. This Place is in the Lat. of 20 deg. 21 min. but in the Draught that I had of this Coast, which was *Tasman*'s, it was laid down in 19 deg. 50 min. and the Shore is laid down as all along joining in one Body or Continent, with some Openings appearing like Rivers; and not

An. 1699.

An. 1699. like Islands, as really they are. See several Sights of it, Table IV. N°. 8, 9, 10. This Place therefore lies more Northerly by 40 min. than is laid down in Mr. *Tasman*'s Draught: And beside its being made a firm, continued Land, only with some Openings like the Mouths of Rivers, I found the Soundings also different from what the prickt Line of his Course shews them, and generally shallower than he makes them: which inclines me to think that he came not so near the Shore as his Line shews, and so had deeper Soundings, and could not so well distinguish the Islands. His Meridian or Difference of Longitude from *Sharks-Bay* agrees well enough with my Account, which is 232 Leagues tho' we differ in Lat. And to confirm my Conjecture that the Line of his Course is made too near the Shore, at least not far to the East of this place, the VVater is there so shallow that he could not come there so nigh.

But to proceed; in the Night we had a small Land-breeze, and in the Morning I weighed Anchor, designing to run in among the Islands, for they had large Channels between them, of a League wide at least, and some 2 or 3 Leagues wide. I sent in my Boat before to sound, and if they found Shole-water to return again; but if they found Water enough, to go ashore

shore on one of the Islands, and stay till the Ship came in; where they might in the mean time search for Water. So we followed after with the Ship, sounding as we went in, and had 20 Fathom, till within 2 Leagues of the Bluff-head, and then we had shole Water, and very uncertain Soundings: Yet we ran in still with an easie Sail, sounding and looking out well, for this was dangerous Work. When we came abreast of the Bluff-head; and about 2 Mile from it, we had but 7 Fathom: Then we Edged away from it, but had no more Water; and running in a little farther, we had but 4 Fathoms: So we Anchored immediately; and yet when we had veered out a third of a Cable we had 7 Fathom Water again; so uncertain was the Water. My Boat came immediately aboard, and told me that the Island was very Rocky and Dry, and they had little hopes of finding Water there. I sent them to sound, and bad them, if they found a Channel of 8 or 10 Fathom Water, to keep on, and we would follow with the Ship. We were now about 4 Leagues within the outer small Rocky Islands, but still could see nothing but Islands within us; some 5 or 6 Leagues long, others not above a Mile round. The large Islands were pretty high; but all appeared Dry, and mostly Rocky and Barren. The Rocks look'd of

a

*An.*1699. a rusty yellow Colour, and therefore I dispair'd of getting Water on any of them: but was in some hopes of finding a Channel to run in beyond all these Islands, could I have spent time here, and either get to the Main of *New Holland*, or find out some other Islands that might afford us Water and other Refreshments: Besides, that among so many Islands, we might have found some sort of Rich Mineral, or Ambergreese, it being a good Latitude for both these. But we had not Sailed above a League farther before our Water grew sholer again, and then we Anchored in 6 Fathom hard Sand.

We were now on the inner side of the Island, on whose outside is the Bluff-point. We rode a League from the Island, and I presently went ashore, and carried Shovels to dig for Water, but found none. There grow here 2 or 3 sorts of Shrubs, one just like Rosemary; and therefore I call'd this *Rosemary* Island. It grew in great plenty here, but had no smell. Some of the other Shrubs had blue and yellow Flowers; and we found 2 sorts of Grain like Beans: The one grew on Bushes; the other on a sort of a creeping Vine that runs along on the Ground, having very thick broad Leaves, and the Blossom like a Bean Blossom, but much larger, and of a deep red Colour, looking very Beautiful. We saw here

here some Cormorants, Gulls, Crabcatchers, *&c.* a few small Land Birds, and a sort of white Parrots, which flew a great many together. We found some Shell-fish, *viz.* Limpits, Perriwinkles, and abundance of small Oysters growing on the Rocks, which were very sweet. In the Sea we saw some green Turtle, a pretty many Sharks, and abundance of Water-Snakes of several sorts and sizes. The Stones were all of rusty Colour, and Ponderous.

We saw a Smoak on an Island 3 or 4 Leagues off; and here also the Bushes had been burned, but we found no other sign of Inhabitants: 'Twas probable that on the Island where the Smoke was there were Inhabitants, and fresh Water for them. In the Evening I went aboard, and consulted with my Officers whether it was best to send thither, or to search among any other of these Islands with my Boat; or else go from hence, and Coast along Shore with the Ship, till we could find some better Place than this was to ride in, where we had shole Water, and lay expos'd to Winds and Tides. They all agreed to go from hence; so I gave Orders to weigh in the Morning as soon as it should be light, and to get out with the Land-breeze.

Accordingly, *August* the 23d. at 5 in the Morning we ran out, having a pretty

An. 1699. fresh Land-breeze at S. S. E. By 8 a Clock we were got out: and very seasonably; for before 9 the Sea-breeze came on us very strong, and increasing, we took in our Topsails and stood off under 2 Courses and a Mizan, this being as much Sail as we could carry. The Sky was clear, there being not one Cloud to be seen; but the Horizon appeared very hazy, and the Sun at setting the Night before, and this Morning at rising, appeared very Red. The Wind continued very strong till Twelve, then it began to abate: I have seldom met with a stronger Breeze. These strong Sea-breezes lasted thus in their Turns 3 or 4 Days. They sprung up with the Sun rise: By 9 a Clock they were very strong, and so continued till Noon, when they began to abate: And by Sun-set there was little Wind, or a Calm till the Land-breezes came; which we should certainly have in the Morning about 1 or 2 a Clock. The Land-breezes were between the S. S. W. and S. S. E. The Sea-breezes between the E. N. E. and N. N. E. In the Night while Calm we fish'd with Hook and Line, and caught good store of Fish, *viz.* Snappers, Breams, Old Wives, and Dog-fish. When these last came we seldom caught any others; for if they did not drive away the other Fish, yet they would be sure to keep them from taking our Hooks, for they would

first

Plate 1.

A Fish taken on the Coast of New Holland.

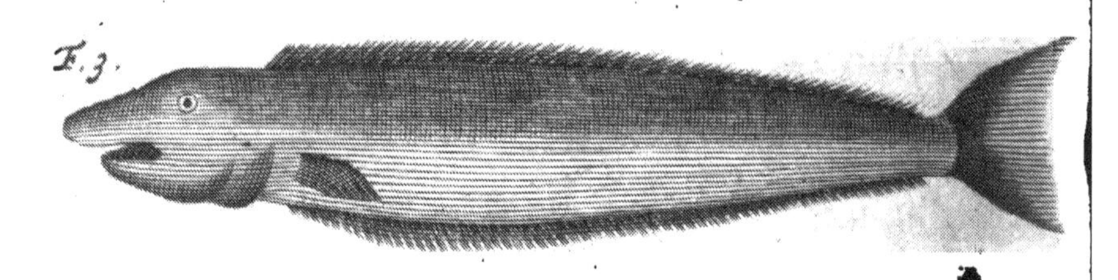

A Cuttle taken near N. Holland.

F. 8.

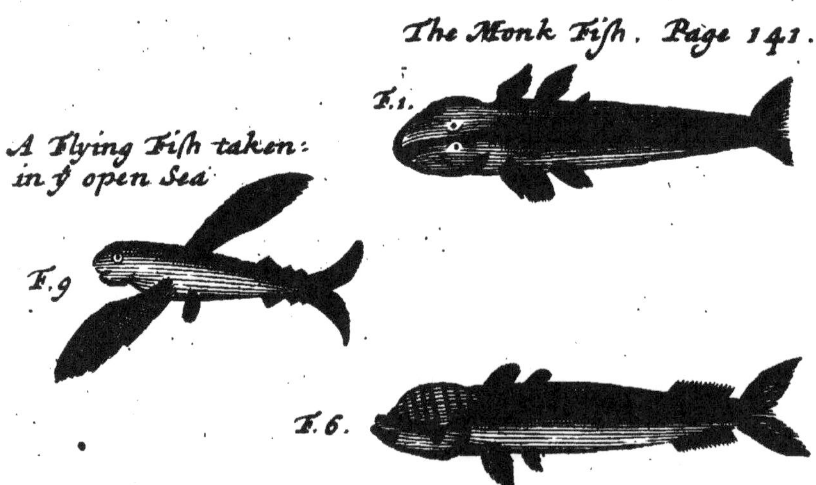

The Monk Fish. Page 141.

A Flying Fish taken in ŷ open Sea

A Remora taken sticking to Sharks backs.

first shave them themselves, biting very greedily. We caught also a Monk-fish, of which I brought home the Picture. See *Fish*, Fig. I.

On the 25th of *August*, we still Coasted along Shore, that we might the better see any Opening; kept sounding, and had about 20 Fathom clean Sand. The 26th Day, being about 4 Leagues off Shore the Water began gradually to sholden from 20 to 14 Fathom. I was Edging in a little towards the Land, thinking to have Anchored: But presently after the Water decreased almost at once, till we had but 5 Fathom. I durst therefore adventure no farther, but steered out the same way that we came in; and in a short time had 10 Fathom (being then about 4 Leagues and a half from the Shore) and even Soundings. I steered away E. N. E. Coasting along as the Land lies. This Day the *Sea-breezes* began to be very moderate again, and we made the best of our way along Shore, only in the Night Edging off a little for fear of Sholes. Ever since we left *Sharks-Bay* we had had fair clear Weather, and so for a great while still.

The 27th Day, we had 20 Fathom Water all Night, yet we could not see Land till 1 in the Afternoon from our Topmast-head. By 3 we could just discern Land from our Quarter-deck: We had then 16 Fathom.

An. 1699. Fathom. The Wind was at N. and we steered E. by N. which is but one point in on the Land: Yet we decreased our Water very fast; for at 4 we had but 9 Fathom; the next Cast but 7, which frighted us; and we then tackt instantly and stood off: But in a short time the Wind coming at N. W. and W. N. W. we tackt again, and steered N. N. E. and then deepned our Water again, and had all Night from 15 to 20 Fathom.

The 28th Day we had between 20 and 40 Fathom. We saw no Land this Day, but saw a great many Snakes, and some Whales. We saw also some *Boobies*, and Noddy-birds; and in the Night caught one of these last. It was of another Shape and Colour than any I had seen before. It had a small long Bill, as all of them have, flat Feet like Ducks Feet; its Tail forked like a Swallow, but longer and broader, and the Fork deeper than that of the Swallow, with very long Wings: The Top or Crown of the Head of this Noddy was Coal-black, having also small black Streaks round about and close to the Eyes; and round these Streaks on each side, a pretty broad white Circle. The Breast, Belly, and under part of the Wings of this Noddy were white: And the Back and upper part of its Wings of a faint black or smoak Colour. See a Picture of this, and of the Com-

Common one, *Birds*, *Fig.* 5, 6. Noddies
are seen in most Places between the *Tropicks*, as well in the *East-Indies*, and on the Coast of *Brazil*, as in the *West-Indies*. They rest a Shore a Nights, and therefore we never see them far at Sea, not above 20 or 30 Leagues, unless driven off in a Storm. When they come about a Ship they commonly perch in the Night, and will sit still till they are taken by the Seamen. They Build on Cliffs against the Sea, or Rocks, as I have said Vol. I. p. 53.

The 30th Day being in Lat. 18 deg. 21 min. we made the Land again, and saw many great Smoaks near the Shore; and having fair Weather and moderate Breezes, I steered in towards it. At 4 in the Afternoon I Anchored in 8 Fathom Water, clear Sand, about 3 Leagues and a half from the Shore. I presently sent my Boat to Sound nearer in, and they found 10 Fathom about a Mile farther in: and from thence still farther in the Water decreased gradually to 9, 8, 7. and at 2 Mile distance to 6 Fathom. This Evening we saw an Eclipse of the Moon, but it was abating before the Moon appear'd to us; for the Horizon was very hazy, so that we could not see the Moon till she had been half an hour above the Horizon: and at two hours, 22 min. after Sun-set, by the reckoning of our Glasses, the Eclipse was quite

gone,

An. 1699. gone, which was not of many Digits. The Moon's Center was then 33 deg. 40 min. high.

The 31st of *August* betimes in the Morning I went ashore with 10 or 11 Men to search for Water. We went armed with Muskets and Cutlasses for our Defence, expecting to see People there; and carried also Shovels and Pickaxes to dig Wells. When we came near the Shore we saw 3 tall black naked Men on the sandy Bay ahead of us: But as we row'd in, they went away. When we were landed I sent the Boat with two Men in her to ly a little from the Shore at an Anchor, to prevent being seiz'd; while the rest of us went after the 3 black Men, who were now got on the top of a small Hill about a quarter of a Mile from us, with 8 or 9 Men more in their Company. They seeing us coming, ran away. When we came on the top of the Hill where they first stood, we saw a plain Savannah, about half a mile from us, farther in from the Sea. There were several Things like Hay-cocks, standing in the Savannah; which at a distance we thought were Houses, looking just like the *Hottentot*'s Houses at the *Cape of G. Hope:* but we found them to be so many Rocks. We searched about these for Water, but could find none, nor any Houses; nor People, for they were all gone.

Then we return'd again to the Place where we landed, and there we dug for Water.

An. 1699.

While we were at work there came 9 or 10 of the Natives to a ſmall Hill a little way from us, and ſtood there menacing and threatning of us, and making a great Noiſe. At laſt one of them came towards us, and the reſt followed at a diſtance. I went out to meet him, and came within 50 yards of him, making to him all the Signs of Peace and Friendſhip I could; but then he ran away, neither would they any of them ſtay for us to come nigh them; for we tried two or three times. At laſt I took two Men with me, and went in the Afternoon along by the Sea-ſide, purpoſely to catch one of them, if I could, of whom I might learn where they got their freſh Water. There were 10 or 12 of the Natives a little way off, who ſeeing us three going away from the reſt of our Men, followed us at a diſtance. I thought they would follow us: but there being for awhile a Sand-bank between us and them, that they could not then ſee us, we made a halt, and hid our ſelves in a bending of the Sand-bank. They knew we muſt be thereabouts, and being 3 or 4 times our Number, thought to ſeize us. So they diſpers'd themſelves, ſome going to the Sea-ſhore, and others beating about

An. 1699. the Sand-hills. We knew by what Rencounter we had had with them in the Morning that we could easily out-run them: so a nimble young Man that was with me, seeing some of them near, ran towards them; and they for some time, ran away before him. But he soon overtaking them, they fac'd about and fought him. He had a Cutlass, and they had Wooden Lances: with which, being many of them, they were too hard for him. When he first ran towards them I chas'd two more that were by the Shore: but fearing how it might be with my young Man, I turn'd back quickly, and went up to the top of a Sand-hill, whence I saw him near me, closely engag'd with them. Upon their seeing me, one of them threw a Lance at me, that narrowly misst me. I discharg'd my Gun to scare them, but avoided shooting any of them: till finding the young Man in great danger from them, and my self in some; and that tho' the Gun had a little frighted them at first, yet they had soon learnt to despise it, tossing up their Hands, and crying *Pooh, Pooh, Pooh*; and coming on afresh with a great Noise, I thought it high time to charge again, and shoot one of them, which I did. The rest, seeing him fall, made a stand again; and my young Man took the opportunity to disengage himself, and come

off

off to me: my other Man alſo was with me, who had done nothing all this while, having come out unarm'd; and I return'd back with my Men, deſigning to attempt the Natives no farther, being very ſorry for what had happen'd already. They took up their wounded Companion: and my young Man, who had been ſtruck through the Cheek by one of their Lances, was afraid it had been poiſon'd: but I did not think that likely. His Wound was very painful to him, being made with a blunt Weapon: but he ſoon recover'd of it.

Among the *N. Hollanders*, whom we were thus engag'd with, there was one who by his Appearance and Carriage, as well in the Morning as this Afternoon, ſeem'd to be the Chief of them, and a kind of Prince or Captain among them. He was a young briſk Man, not very tall, nor ſo perſonable as ſome of the reſt, tho' more active and couragious: He was painted (which none of the reſt were at all) with a Circle of white Paſte or Pigment (a ſort Lime, as we thought) about his Eyes, and a white ſtreak down his Noſe from his Forehead to the tip of it. And his Breaſt and ſome part of his Arms were alſo made white with the ſame Paint: not for Beauty or Ornament, one would think, but as ſome wild *Indian* Warriors are ſaid to do, he

An. 1699. seem'd thereby to design the looking more terrible; this his Painting adding very much to his natural Deformity; for they all of them of the most unpleasant Looks and the worst Features of any People that ever I saw, tho' I have seen great variety of Savages. These *N. Hollanders* were probably the same sort of People as those I met with on this Coast in my *Voyage round the World*; [See Vol. I. p. 464, *&c.*] for the Place I then touch'd at was not above 40 or 50 Leagues to the N. E. of this: And these were much the same blinking Creatures (here being also abundance of the same kind of Flesh-flies teizing them) and with the same black Skins, and Hair frizled, tall and thin, *&c.* as those were: But we had not the opportunity to see whether these, as the former, wanted two of their fore-Teeth.

We saw a great many places where they had made Fires; and where there were commonly 3 or 4 Boughs stuck up to Windward of them; for the Wind (which is the Sea-breeze) in the day-time blows always one way with them; and the Land-breeze is but small. By their Fire-places we should always find great heaps of Fish-shells, of several sorts; and 'tis probable that these poor Creatures here lived chiefly on the Shell-fish, as those I before describ'd did on small Fish, which they caught in

Wires

Wires or Holes in the Sand at Low-water. These gather'd their Shell-fish on the Rocks at Low-water; but had no Wires (that we saw) whereby to get any other sorts of Fish: As among the former I saw not any heaps of Shells as here, though I know they also gather'd some Shell-fish. The Lances also of those were such as these had; however they being upon an Island, with their Women and Children, and all in our Power, they did not there use them against us, as here on the Continent, where we saw none but some of the Men under Head, who come out purposely to observe us. We saw no Houses at either Place; and I believe they have none, since the former People on the Island had none, tho' they had all their Families with them.

An. 1699.

Upon returning to my Men I saw that tho' they had dug 8 or 9 Foot deep, yet found no Water. So I returned aboard that Evening, and the next day, being *September* 1st, I sent my Boatswain ashore to dig deeper, and sent the Sain with him to catch Fish. While I staid aboard I observed the flowing of the Tide, which runs very swift here, so that our Nun-buoy would not bear above the Water to be seen. It flows here (as on that part of *N. Holland* I describ'd formerly, about 5 Fathom: and here the Flood runs S. E. by S. till the last Quarter; then it sets

 right

An. 1699. right in towards the Shore (which lies here S. S. W. and N. N. E.) and the Ebb runs N. W. by N. When the Tides slackned we Fish'd with Hook and Line, as we had already done in several Places on this Coast; on which in this Voyage hitherto, we had found but little Tides: but by the Heighth, and Strength, and Course of them hereabouts, it should seem that if there be such a Passage or Streight going through Eastward to the Great *South Sea*, as I said one might suspect, one would expect to find the Mouth of it somewhere between this Place and *Rosemary* Island, which was the part of *N. Holland* I come last from.

Next Morning my Men came aboard and brought a Rundlet of brackish Water which they got out of another Well that they dug in a Place a mile off, and about half as far from the Shore; but this Water was not fit to drink. However we all concluded that it would serve to boil our Oatmeal, for Burgoo, whereby we might save the Remains of our other Water for drinking, till we should get more; and accordingly the next day we brought aboard 4 Hogsheads of it: but while we were at work about the Well we were sadly pester'd with the Flies, which were more troublesome to us than the Sun, tho' it shone clear and strong upon us all the while,

very

very hot. All this while we ſaw no more of the Natives, but ſaw ſome of the Smoaks of ſome of their Fires at 2 or 3 miles diſtance.

The Land hereabouts was much like that part of *New Holland* that I formerly deſcribed [Vol. I. p. 463.] 'tis low, but ſeemingly barricado'd with a long Chain of Sand-hills to the Sea, that let's nothing be ſeen of what is farther within Land. At high Water the Tides riſing ſo high as they do, the Coaſt ſhews very low: but when 'tis low Water it ſeems to be of an indifferent heighth. At low Water-Mark the Shore is all Rocky, ſo that then there is no Landing with a Boat; but at high Water a Boat may come in over thoſe Rocks to the Sandy Bay, which runs all along on this Coaſt. The Land by the Sea for about 5 or 600 yards is a dry Sandy Soil, bearing only Shrubs and Buſhes of divers ſorts. Some of theſe had them at this time of the year, yellow Flowers or Bloſſoms, ſome blue, and ſome white; moſt of them of a very fragrant Smell. Some had Fruit like Peaſecods; in each of which there were juſt ten ſmall Peas: I opened many of them, and found no more nor leſs. There are alſo here ſome of that ſort of Bean which I ſaw at *Roſemary-Iſland*: and another ſort of ſmall, red, hard Pulſe, growing in Cods alſo, with

An. 1699. little black Eyes like Beans. I know not their Names, but have ſeen them uſed often in the *East-Indies* for weighing Gold; and they make the ſame uſe of them at *Guinea*, as I have heard, where the Women alſo make Bracelets with them to wear about their Arms. Theſe grow on Buſhes: but here are alſo a Fruit like Beans growing on a creeping ſort of Shrub-like Vine. There was great plenty of all theſe ſorts of Cod-fruit growing on the Sand-hills by the Sea-ſide, ſome of them green, ſome ripe, and ſome fallen on the Ground: but I could not perceive that any of them had been gathered by the Natives; and might not probably be wholeſome Food.

The Land farther in, that is lower than what borders on the Sea, was, ſo much as we ſaw of it, very plain and even; partly Savannahs, and partly Woodland. The Savannahs bear a ſort of thin courſe Graſs. The Mould is alſo a courſer Sand than that by the Sea-ſide, and in ſome places 'tis Clay. Here are a great many Rocks in the large Savannah we were in, which are 5 or 6 Foot high, and round at top like a Hay-cock, very remarkable; ſome red, and ſome white. The Woodland lies farther in ſtill; where there were divers ſorts of ſmall Trees, ſcarce any three Foot in circumference; their Bodies 12 or

14 Foot high, with a Head of small Knibs or Boughs. By the sides of the Creeks, especially nigh the Sea, there grow a few small black Mangrove-Trees. An. 1699.

There are but few Land-Animals. I saw some Lizards; and my Men saw two or three Beasts like hungry Wolves, lean like so many Skeletons, being nothing but Skin and Bones: 'Tis probable that it was the Foot of one of those Beasts that I mention'd as seen by us in *N. Holland*, [Vol. I. p. 463.] We saw a Rackoon or two, and one small speckled Snake.

The Land-fowls that we saw here were Crows (just such as ours in *England*) small Hawks, and Kites; a few of each sort: but here are plenty of small Turtle-Doves, that are plump, fat and very good Meat. Here are 2 or 3 sorts of smaller Birds, some as big as Larks, some less; but not many of either sort. The Sea-Fowl are Pelicans, Boobies, Noddies, Curlews, Sea-pies, *&c.* and but few of these neither.

The Sea is plentifully stock'd with the largest Whales that I ever saw: but not to compare with the vast ones of the *Northern* Seas. We saw also a great many Green Turtle, but caught none; here being no Place to set a Turtle-Net in; here being no Channel for them, and the Tides running so strong. We saw some Sharks, and

An. 1699. and Paracoots; and with Hooks and Lines we caught some Rock-fish and Old Wives. Of Shell-fish, here were Oysters both of the common kind for Eating, and of the Pearl-kind: and also Wilks, Conchs, Muscles, Limpits, Perriwinkles, *&c.* and I gather'd a few strange Shells; chiefly a sort not large, and thick-set all about with Rays or Spikes growing in Rows.

And thus having ranged about, a considerable time, upon this Coast, without finding any good fresh Water, or any convenient Place to clean the Ship, as I had hop'd for: And it being moreover the heighth of the dry Season, and my Men growing Scorbutick for want of Refreshments, so that I had little Incouragement to search further; I resolved to leave this Coast, and accordingly in the beginning of *September* set Sail towards *Timor*.

AN

Tab. 1.

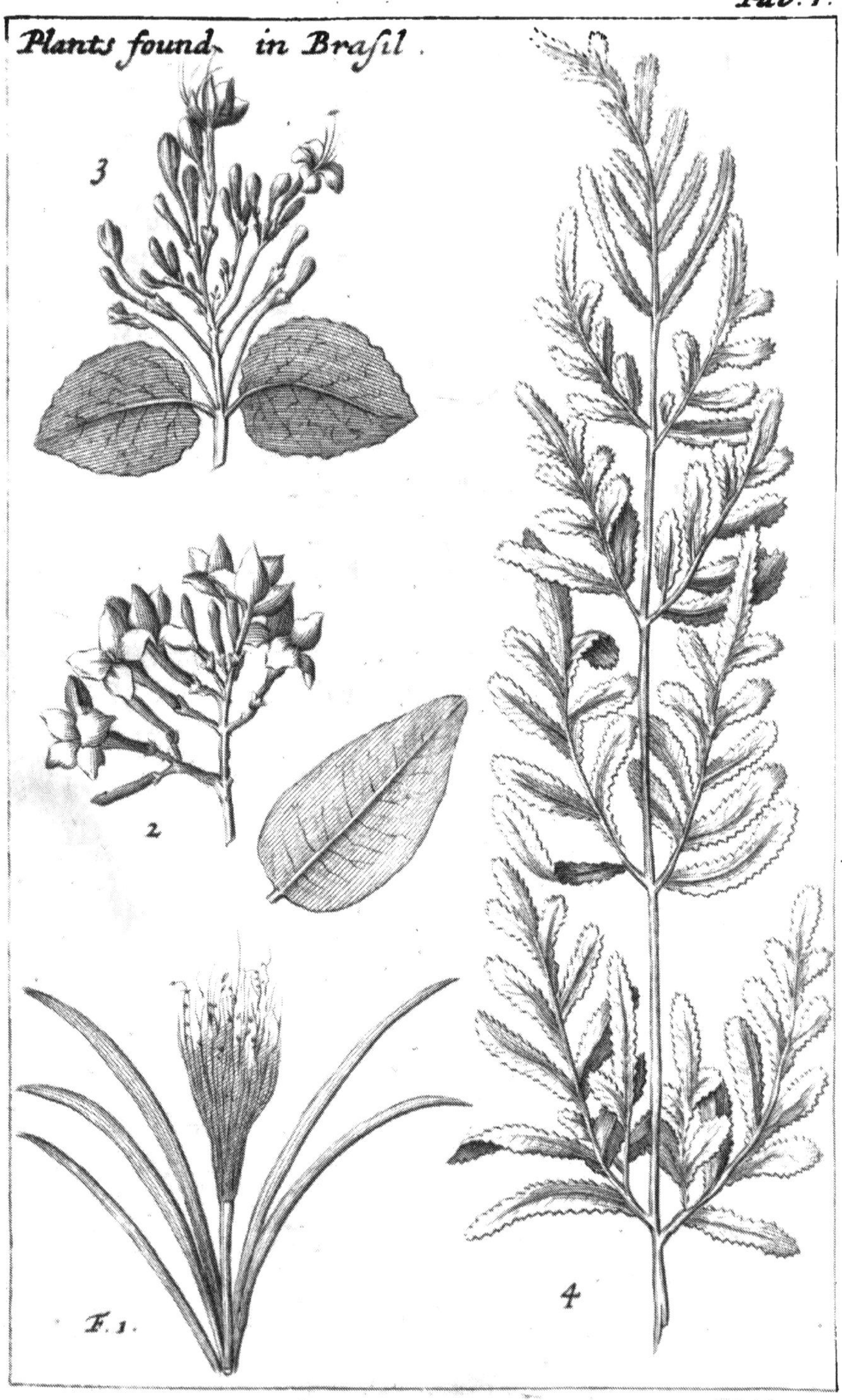

AN ACCOUNT

Of ſeveral

PLANTS

Collected in

Braſil, *New Holland*, *Timor*, and *New Guinea*, referring to the Figures Engraven on the Copper Plates.

TAB. 1. Fig. 1. *Cotton-flower* from *Baya* in *Braſil*. The Flower conſiſts of a great many Filaments, almoſt as ſmall as Hairs, betwixt 3 and 4 Inches long, of a Murrey-colour; on the top of them ſtand ſmall aſh-colour'd *apices*. The pedicule of the Flower is inclos'd at the bottom with five narrow ſtiff Leaves, about ſix Inches long. There is one of this *genus* in Mr. *Ray*'s Supplement, which agrees

grees exactly with this in every respect, only that is twice larger at the least. It was sent from *Surinam* by the Name of *Momoo.*

Tab. 1. Fig. 2. *Jasminum Brasilianum luteum, mali limoniæ folio nervoso, petalis crassis.*

Tab 1. Fig. 3. *Crista Pavonis Brasiliana Bardanæ foliis.* The Leaves are very tender and like the top Leaves of *Bardana major*, both as to shape and texture: In the Figure they are represented too stiff and too much serrated.

Tab. 1. Fig. 4. *Filix Brasiliana Osmundæ minori serrato folio.* This Fern is of that kind, which bears it's Seed-Vessels in Lines on the edge of the Leaves.

Tab. 2. Fig 1. *Rapuntium Novæ Hollandiæ, flore magno coccineo.* The *Perianthium* compos'd of five long pointed Parts, the Form of the Seed-Vessel and the smalness of the Seeds, together with the irregular shape of the Flower and thinness of the Leaves, argue this Plant to be a *Rapuntium.*

Tab. 2. Fig. 2. *Fucus foliis capillaceis brevissimis, vesiculis minimis donatis.* This elegant *fucus* is of the *Erica Marina* or *Sargazo* kind, but has much finer parts than that. It was collected on the Coast of *New Holland.*

Tab.

Plants found in New Holland.
F. 1.
2.
3.
4.

Tab. 3.
Plants found in New Holland.
F. 1.
2
3
4

Tab. 2. Fig. 2. *Ricinoides Novæ Hollandæ angulosо crasso folio.* This Plant is shrubby, has thick woolly Leaves, especially on the under side. Its Fruit is tricoccous, hoary on the out-side with a *Calix* divided into five parts. It comes near *Ricini fructu parvo frutosa Curassavica, folio Phylli, P. B. pr.*

Tab. 2. Fig. 2. *Solanum spinosum Novæ Hollandiæ Phylli foliis subrotundis.* This new *Solanum* bears a blewish Flower like the others of the same Tribe; the Leaves are of a whitish colour, thick and woolly on both sides, scarce an Inch long and near as broad. The Thorns are very sharp and thick set, of a deep Orange colour, especially towards the Points.

Tab. 3. Fig. 1. *Scabiosa (forte) Novæ Hollandiæ, statices foliis subtus argenteis.* The Flower stands on a Foot-stalk four Inches long, included in a rough Calix of a yellowish colour. The Leaves are not above an Inch long, very narrow like *Thrift*, green on the upper and hoary on the under side, growing in tufts. Whether this Plant be a *Scabious*, *Thrift* or *Helichrysum* is hard to judge from the imperfect Flower of the dry'd Specimen.

Tab. 3. Fig. 2. *Alcea Novæ Hollandiæ foliis augustis utrinque villosis.* The Leaves stalk and under side of the Perianthium of this Plant are all woolly. The Petala are very

very tender, five in number, ſcarce ſo large as the Calix: In the middle ſtands a a *Columella* thick ſet with thrummy *apiculæ*, which argue this Plant to belong to the Malvaceous kind.

Tab. 3. Fig. 3. Of what *genus* this Shrub or Tree is, is uncertain, agreeing with none yet deſcrib'd, as far as can be judg'd, by the State it is in. It has a very beautiful Flower, of a red colour as far as can be gueſs'd by the dry *Specimen*, conſiſting of ten large *Petala*, hoary on both ſides, eſpecially underneath; the middle of the Flower is thick ſet with *Stamina*, which are woolly at the bottom, the length of the *Petala*, each of them crown'd with its *Apex*. The *Calix* is divided into five round pointed parts. The Leaves are like thoſe of *Amelanchier Lob.* green a top and very woolly underneath, not running to a point, as is common in others, but with an Indenture at the upper end.

Tab. 3. Fig. 4. *Dammara ex Nova Hollandia, Sanamunda ſecundæ Clyſii foliis.* This new *genus* was firſt ſent from *Amboyna* by Mr. *Rumphius*, by the Name of *Dammara*, of which he tranſmitted two kinds; one with narrow and long ſtiff Leaves, the other with ſhorter and broader. The firſt of them is mention'd in Mr. *Petiver*'s *Centuria*, p. 350. by the Name of *Arbor hortenſis Javanorum foliis*

viſci

Plants found in New Holland & Timor.
F.1.
2.
3.
4.

visce angustioribus aromaticis floribus, spicatis stamineis lutescentibus; Muſ. Pet. As alſo in Mr. *Ray*'s Supplement to his Hiſtory of Plants now in the Preſs. This is of the ſame *genus* with them, agreeing both in Flower and Fruit, tho' very much differing in Leaves. The Flowers are ſtamineous and ſeem to be of an herbaceous colour, growing among the Leaves, which are ſhort and almoſt round, very ſtiff and ribb'd on the under ſide, of a dark green above, and a pale colour underneath, thick ſet on by pairs, anſwering one another croſs-ways, ſo that they cover the Stalk. The Fruit is as big as a Pepper-corn, almoſt round, of a whitiſh colour, dry and tough, with a Hole on the top, containing ſmall Seeds. Any one that ſees this Plant without it's Seed-Veſſels, would take it for an *Erica* or *Sanamunda*. The Leaves of this Plant are of a very aromatick Taſt.

Tab. 4. Fig. 1. *Equiſetum Novæ Hollandiæ fruteſcens foliis longiſſimis*. 'Tis doubtful whether this be an *Equiſetum* or not; the texture of the Leaves agrees beſt with that *genus* of any, being articulated one within another at each Joint, which is only proper to this Tribe. The longeſt of them are about nine Inches.

Tab. 4. Fig. 2. *Colutea Novæ Hollandiæ floribus amplis coccineis, umbellatim diſpoſitis macula purpurea notatis.* There being no Leaves to this Plant, 'tis hard to ſay what *genus*

genus it properly belongs to. The Flowers are very like to the *Colutea Barba Jovis folio flore coccineo Breynii*; of the ſame Scarlet colour, with a large deep purple Spot in the *vexillum*, but much bigger, coming all from the ſame point after the manner of an Umbel. The rudiment of the pod is very woolly, and terminates in a Filament near two Inches long.

Tab. 4. Fig. 3. *Conyza Nova Hollandia anguſtis Roriſmarini foliis.* This Plant is very much branch'd and ſeems to be woody. The Flowers ſtand on very ſhort Pedicules, ariſing from the *ſinus* of the Leaves, which are exactly like *Roſemary*, only leſs. It taſts very bitter now dry.

Tab. 4. Fig. 4. *Moboh Inſulæ Timor.* This is a very odd Plant, agreeing with no deſcrib'd *genus*. The Leaf is almoſt round, green on the upper ſide and whitiſh underneath, with ſeveral Fibres running from the inſertion of the Pedicule towards the circumference 'tis umbilicated as *Cotyledon aquatica* and *Faba Ægyptia.* The Flowers are white ſtanding on ſingle Foot-ſtalks, of the ſhape of a *Stramonium*, but divided into four points only, as is the *Perianthium*.

Tab. 5. Fig. 1. *Fucus ex Nova Guinea uva marina dictus, foliis variis.* This beautiful *fucus* is thick ſet with very ſmall ſhort tufts of Leaves, which by the help

of

Tab
Plants found in yᵉ Sea neer New Guinea.
F. 1.
2.

of a magnifying Glaſs, ſeem to be round and articulated, as if they were Seed-Veſſels; beſides theſe, there are other broad Leaves, chiefly at the extremity of the Branches, ſerrated on the edges. The *veſiculæ* are round, of the bigneſs expreſs'd in the Figure.

Tab. 5. Fig. 2. *Fucus ex Nova Guinea Fluviatilis Piſanæ J. B. foliis.* Theſe Plants are ſo apt to vary in their Leaves, according to their different States, that 'tis hard to ſay this is diſtinct from the laſt. It has in ſeveral Places (not all expreſs'd in the Figure) ſome of the ſmall ſhort Leaves, or Seed-Veſſels mention'd in the former; which makes me apt to believe it the ſame, gather'd in a different ſtate; beſides the broad Leaves of that and this agree as to their Shape and Indentures.

An Account of ſome Fiſhes that are Figured *in* Plate 2. & 3.

See Plate 3. Fig. 5.

THis is a Fiſh of the Tunny-kind, and agrees well enough with the Figure in Tab. 3. of the Appendix to Mr. *Willughby*'s Hiſtory of Fiſhes under the Name of *Gurabuca* ; it differs ſomething, in the Fins eſpecially, from *Piſo*'s Figure of the *Guarapucu.*

See Plate 3. Figure 4.

This reſembles the Figure of the *Guaperva maxima candata* in *Willughby*'s *Ichthyol.* Tab. 9, 23. and the *Guaperva* of *Piſo*, but does not anſwer their Figures in every particular.

See Plate 2. Figure 2.

There are 2 ſorts of *Porpuſſes:* The one the long-ſnouted *Porpuſs*, as the Seamen call it; and this is the *Dolphin* of the *Greeks.* The other is the Bottle-noſe *Porpuſs*, which is generally thought to be the *Phæcena* of *Ariſtotle.*

Plate 2. Figure 7.

This is the *Guaracapema* of *Piſo* and *Marcgrave*, by others call'd the *Dorado.* 'Tis Figured in *Willughby*'s *Ichthyol.* Tab. O. 2. under the Name of *Delphin Belgis.*

THE

A Dolphin as it is usually called by our seamen, taken in the open Sea.

Plate 3.

Fish of the Tunny kind taken on ye Coast of N. Holland

F. 5.

A Fish called by the seamen the Old Wife.

F. 4.

THE

INDEX.

A.

B.

Clocking-

D.

F.

G.

H.

J.

Jeni-

N.

O.

P.

A

W.

Y.

F I N I S.

A Catalogue of Books.

A Relation of two several Voyages made into the *East-Indies*, by *Christopher Fryke*, Surg. and *Christopher Schewitzer*. The whole containing an Exact Account of the *Customs*, *Dispositions*, *Manners*, *Religion*, &c. of the several *Kingdoms* and *Dominions* in those parts of the World in General: But in a more particular manner, describing those Countries which are under the Power and Government of the Dutch. *Octavo* Price 4 *s*.

Discourses on the Publick Revenues. and on the Trade of *England*. In Two Parts, *viz*. I. Of the Use of Political Arithmetick, in all Considerations about the Revenues and Trade. II. On Credit, and the Means and Methods by which it may be restor'd. III. On the Management of the King's Revenues. IV. Whether to Farm the Revenues, may not, in this Juncture, be most for the Publick Service? V. On the Publick Debts and Engagements. Part I. To which is added, A Discourse upon Improving the Revenue of the State of *Athens*. Written Originally in *Greek*; and now made *English* from the Original, with some Historical Notes.

Discourses on the *Publick Revenues*, and on the Trade of *England*; which more immediately Treat of the Foreign Traffick of this Kingdom. *viz*. I. That Foreign Trade is beneficial to *England*. II. On the Protection and Care of Trade. III. On the Plantation Trade. IV. On the *East-India* Trade. Part II. To which is added the late *Essay* on the *East-India* Trade.

An Essay upon the Probable Methods of making a *People Gainers*, in the Balance of Trade. Treating of these Heads; *viz*. Of the People of *England*. Of the Land of *England*, and its Product. Of our Payments to the Publick, and in what manner the Balance of Trade may be thereby effected. That a Country cannot increase in Wealth and Power, but by private Men doing their Duty to the Publick, and but by a steady Course of Honesty and Wisdom, in such as are Trusted with the Administration of Affairs.

A Discourse upon *Grants* and *Resumptions*. Shewing how our Ancestors have proceeded with such Ministers as have procured to themselves Grants of the Crown-Revenue; and that the forfeited Estates in *Ireland* ought to be applied towards the Payment of the Publick Debts.

Essays upon I. The Balance of Power. II. The Right of making War, Peace and Alliances. III. Universal Monarchy. To which is added, an APPENDIX containing the Records referr'd to in the Second Essay. These five by the Author of, *The Essays on Ways and Means*.

Several Discourses, Concerning the Shortness of Humane Charity. The Perfection of the Mercy of God. The Difference of Times with respect to Religion. The Joy which the Righteous have in God. The Secret Blasting of Men. The Instructive Discipline of God. The Danger of Unfaithfulness to God. The Malignity of Popery. The Deceitfulness of Sin. The Conversion of a Sinner. Also, the Prayer used before Sermon. Vol. I. The 2d Edit. Pr. 5 *s*. Se-

——Several Discourses, concerning the true Valuation of Man. The Necessary Repentance of a Sinner. The Exercise and Progress of a Christian. The Frailty of Humane Nature. The Justice of one towards another. The Nature of Salvation by Christ, &c. Being Twenty Sermons. Vol. II. Both by the Reverend and Learned *Benjamin Whichcote*, sometime Minister of St. *Lawrence Jury*, *London*. Examined and Corrected by his own Notes; and Published by *John Jeffery*, D. D. Archdeacon of *Norwich*, Price 5 *s*.

Three Practical Essays, *viz*. On *Baptism*, *Confirmation*, and *Repentance*. Containing Instructions for a Holy Life: With earnest Exhortations, especially to young Persons, drawn from the Considerations of the Severity of the Discipline of the Primitive Church. By *Samuel Clarke*, M. A. Chaplain to the Right Reverend Father in God *John* Lord Bishop of *Norwich*: Price 3 *s*.

A Paraphrase on the Gospels of St. *Matthew*, St. *Mark*, St. *Luke* and St. *John*. In Two Volumes. Written by *Samuel Clarke*, A. M. Chaplain to the Right Reverend Father in God *John* Lord Bishop of *Norwich*. 8*vo*.

Jacobi Rohaulti Physica. Latine vertit, recensuit, & uberioribus jam Annotationibus, ex illustrissimi *Isaaici Newtoni* Philosophia maximam partem haustis, amplificavit & ornavit *Samuel Clarke*, A. M. Admodum Reverendo in Christo patri, Joanni Episcopo Norvicensi, a Sacris Domesticis. Accedunt etiam in hac Secunda Editionæ, novæ aliquot Tabulæ æri incisæ. 8*vo*.

Confessio, sive Declaratio, Sententiæ Pastorum, qui in Fœderato Belgio remonstrantes vocantur, super præcipuis Articulis Religionis Christianæ. 12*ves*. price 1 *s*. 6 *d*.

Devotions, *viz*. Confessions, Petitions, Intercessions, and Thanksgivings for every Day of the Week; and also Before, At, and After the Sacrament: With Occasional Prayers for all Persons whatsoever. By *Thomas Bennet*, M. A. Rector of St. *James*'s in *Colchester*, and Fellow of St. *John*'s Colledge in *Cambridge*.

The God-Father's Advice to his Son. Shewing the Necessity of Performing the Baptismal Vow, and the Danger of neglecting it With general Instructions to young Persons to lead a Religious Life, and prepare them for their Confirmation. Very necessary for Parents, &c. to give their Children, or others committed to their Care. By *John Birket*, Vicar of *Milford* and *Hordle* in *Hampshire*. The Second Edition, with a Preface, Price 3 *d*. 100 for 20 *s*.

The Government of the Passions, according to the Rules of Reason and Religion, *viz*. Love, Hatred, Desire, Eschewing, Hope, Despair, Fear, Anger, Delight and Sorrow. *Twelves*.

Some Reflections on that part of a Book called *Amyntor*: Or, The Defence of *Milton*'s Life, which relates to the Writings of the Primitive Fathers and the Canon of the New Testament. *In a Letter to a Friend*. Octavo.

A

A Catalogue of Books.

A Treatise of Morality. In Two Parts. Written in *French* by *F. Malbranch*, Author of *The Search after Truth*. And Translated into *English* by *James Shipton*, M. A.

The Memoirs of Monsieur *Pontis*, who served in the *French* Armies 56 Years. Translated by *Charles Cotton* Esq; *Folio*.

Processus integri in Morbis fere omnibus Curandis, a Duo. *Tho. Sydenham* conscripti *Duodecimo*.

Dr. *Sydenham*'s Practice of Physick, Faithfully Translated into *English* with large Annotations, Animadversions, and Practical Observations on the same, By *W. Salmon*, M. D. *Twelves*.

The Penitent, or Entertainments for *Lent*. Written in *French* by R. F. *N. Caussin*, and translated into *English* by Sir *B. B.* Tenth Edition. To which are added several Sculptures.

A New Method of Curing all Sorts of Fevers, without taking any thing by the Mouth. Being a New Prescription for giving the Bark in Clyster. Whereby all the Inconveniences of administring it in any other Form are avoided; and a more speedy, certain Cure is obtained. Writ by *A. Helvet*, M. D. *The Second Edition*.

Mr. *Wingate*'s Arithmetick: Containing a plain and familiar Method for attaining the Knowledge and Practice of Common Arithmetick. *The Tenth Edition, very much enlarged*. By *John Kirsey*, late Teacher of the Mathematicks.

The History of the Inquisition, as it is exercised at *Goa*. Written in *French*, by the Ingenious Monsieur *Dellon*, who laboured five Years under those Severities. With an Account of his Deliverance, Done into *English* by the learned *Henry Wharton*, M. A. Chaplain to his Grace the late Archbishop of *Canterbury*.

The Artificial Clock-Maker. A Treatise of Watch and Clock-work. Wherein the Art of Calculating Numbers for most sorts of Movements is explained, to the Capacity of the Unlearned. Also, the History of Watch and Clock-work, both Ancient and Modern. With other Useful Matters never before Publish'd. *The Second Edition Enlarged*. To which is added a Supplement, containing. 1. The Anatomy of a Watch and Clock. 2. Monsieur *Romer*'s Satellite-Instrument, with Observations concerning the Calculation of the Eclipses of *Jupiter*'s *Satellites*, and to find the *Longitude* by them. 3. A nice way to correct Pendulum Watches. 4. M. *Flamsteed*'s Equation Tables. 5. To find a *Meridian-Line*, for the Governing of Watches, and other Uses. 6. To make a *Telescope* to keep a Watch by the Fixed Stars. By *W. D.* M. A. price 1 *s.* 6 *d.*

Arcana Imperii detecta: Or, divers select Cases in Government; more particularly, Of the Obeying the unjust Commands of a Prince. Of the Renunciation of a Right to a Crown. Of the Proscription of a limitted Prince and his Heirs. Of the Trying, Condemning and Execution of a Crowned Head. Of the Marriage of a Prince and Princess. Of the Detecting of Conspiracies against a Government. Of Subjects Revolting from a Tyranical Prince. Of

Exclu-

Excluding Foreigners from Publick Employments. Of Constituting Extraordinary Magistrates upon Extraordinary Occasions. Of Subjects Anticipating the Execution of Laws. Of Toleration of Religion. Of Peace and War, &c. *With the Debates, Arguments and Resolutions of the greatest Statesmen, in several Ages and Governments thereupon.*

The Royal Dictionary, in Two Parts. I. *French* and *English*. II. *English* and *French*. The *French* taken out of the Dictionaries of *Richelet*, *Furetiere*, *Tachart*, the Great Dictionary of the *French-Academy*, and the Remarks of *Vaugelas*, *Menage*, and *Bouhours*. And the *English* Collected chiefly out of the best Dictionaries, and the Works of the greatest Masters of the *English* Tongue; such as Archbishop *Tillotson*, bishop *Sprat*, Sir *Roger L' Estrange*, Mr. *Dryden*, Sir *William Temple*, &c. For the Use of his Highness the Duke of *Glocester*. By Mr. *Boyer*. *Quarto.*

—— *Idem* in Octavo.

Bennet of Schism Price 2 *s.* 6 *d.*

—— Defence of it pr. 1 *s.*

History of *England*.

Life of K. James pr. 5 *s.*

Life of K. William pr. 6 *s.*

Cambrige Concordance. Folio,

Collier's Essays. Octavo.

Milners Reflections on L' Clerk, Octavo pr. 3 *s.* 6 *d.*

Salmon's Dispensatory. Octavo.

Seneca's Morals. Octavo.

Newcomb's Sermons.

Sherlock's Sermons. Octavo.

Sharp's Sermons Octavo.

Scot's Sermons. 2 Vol. Octavo.

—— Christian Life, in 5 Vol. Octavo.

A View of the Posture of Affairs in *Europe* both in Church and State. I. The Antient Pretensions of the two Families of *Austria* and *Bourbon*, to the *Spanish* Monarchy. II. The Balance of the Power of *Europe*, setled by *Charles* V. and how it came to be broke. III. A View of the Courts of *Europe*, and their present Disposition and State relating to War. IV. Of the State of the Church of *Rome*, and the Decay of the Protestant Interest in *Europe*. Written by a Gentleman by way of Letter.

The Surgeons Assistant. In which is plainly discovered the True Origin of most Diseases. Treating particularly of the Plague, *French* Pox, Leprosie. &c. Of the Biting of mad Dogs, and other Venemous Creatures, Also A Compleat Treatise of Cancers and Gangreens. With an Enquiry whether they have any Alliance with Contagious Diseases. Their most Easie and Speedy Method of Cure. With divers Approved Receipts. By *John Browne*, Sworn Surgeon in Ordinary to his late most Excellent Majesty King *William* III. and late Senior Surgeon of St. *Thomas*'s Hospital in *Southwark*. pr. 2 *s.* 6 *d.*

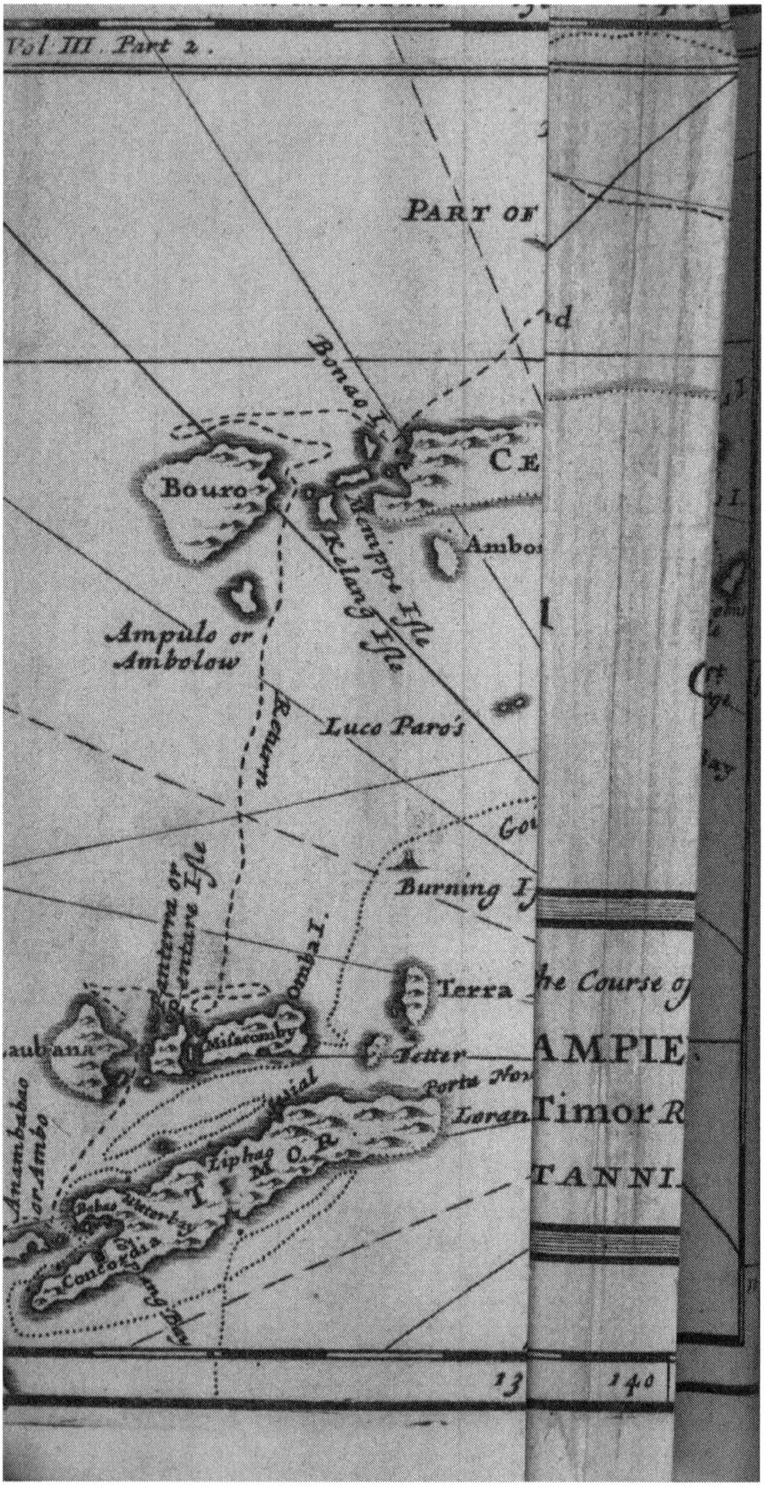

PART OF

Bonao I.

CE

Bouro

Manippe Isle

Kelang Isle

Ambo

Ampulo or Ambolow

Return

Luco Paro's

Burning I

Omba I.

Terra

Tetter

Porto Nov

Loran

Misacomby

Anambabao or Ambo

T I M O R

Liphao

Babao

Concordia

the Course of

AMPIE

Timor R

TANNI

13

140

A CONTINUATION OF A VOYAGE TO *NEW-HOLLAND*, &c. In the Year 1699.

Wherein are described,

The Islands *Timor*, *Rotee* and *Anabao*. A Passage between the Islands *Timor* and *Anabao*. *Copang* and *Laphao* Bays. The Islands *Omba*, *Fetter*, *Bande* and *Bird*. A Description of the Coast of *New-Guinea*. The Islands *Pulo Sabuda*, *Cockle*, King *William's*, *Providence*, *Garret Dennis*, *Ant. Cave's* and St *John's*. Also a new Passage between *N. Guinea* and *Nova Britannia*. The Islands *Ceram*, *Bonao*, *Bouro*, and several Islands before unknown. The Coast of *Java*, and Streights of *Sunda*. Author's Arrival at *Batavia*, *Cape of Good Hope*, St. *Helens*, I. *Ascension*, &c.

Their Inhabitants, Customs, Trade, &c. Harbours, Soil, Birds, Fish, &c. Trees, Plants, Fruits, &c.

Illustrated with Maps and Draughts: Also divers Birds, Fishes, &c. not found in this part of the World, Ingraven on Eighteen Copper-Plates.

By Captain *William Dampier*.

London, Printed by *W. Botham*; for *James Knapton*, at the *Crown* in St *Paul's* Church-Yard. 1709.

THE CONTENTS.

CHAP. I.

 ty,

CHAP. II.

CHAP. III.

William's *Island. A Description of it. Plying on the Coast of* New-Guinea. *Fault of the Draughts.* Providence *Island. They cross the Line. A Snake pursued by Fish. Squally Island. The Main of* New-Guinea.

CHAP. IV.

The Main Land of New Guinea. *Its Inhabitants.* Slingers *Bay. Small Islands.* Garret Dennis *Isle described. Its Inhabitants. Their Proes.* Anthony Caves *Island. Its Inhabitants. Trees full of Worms found in the Sea.* St. Johns *Island. The main Land of* New Guinea. *Its Inhabitants. The Coast described. Cape and Bay* St. George. *Cape* Orford. *Another Bay. The Inhabitants there. A large account of the Author's attempts to Trade with them. He names the place* Port Moun-

 tague.

CHAP.

CHAP. V.

CHAP.

CHAP. VI.

A
CATALOGUE
OF THE
Mapps and *Copper-Plates*,

11. Squally

Islands Laubano *and* Pantara, N. 4.

Numb. I. *Strange Fishes taken on the Coast of* New Guinea.

Numb. II. *A Fish of a pale red, all parts of it, except the Eye*, Fig. 1.

A strange large Batt taken on the Island Pulo Sabuda *in* New Guinea, *described. p.* 199 Fig. 2.

A large Bird. Fig. 3.

Numb. III. *Three strange Birds; one described. p.* 93
Another described. p. 165

Numb. IV. *Several Fishes taken on the Coast of* New Guinea.

Numb. V. *The* Mountain-Cow; *or, as some think, the* Hippopotamus, *described in Capt.* Dampier's 2d *Vol. in* Campeachy, *p.* 102, 3, 4, 5, 6, 7.

BOOKS

BOOKS Printed for J. Knapton, *at the* Crown *in* St. Paul's *Church-Yard.*

AN Historical Geography of the New Testament in two Parts. Part I. *The Journeyings of Our Lord and Saviour* Jesus Christ. Part II. *The Travels and Voyages of St* Paul, &c. Being a *Geographical* and *Historical* Account of all the Places mention'd, or referr'd to, in the *Books of the New Testament*; Very useful for understanding the History of the said Books, and several Particular Texts. To which end there is also added a Chronological Table. Throughout is inserted the *Present* State of such Places, as have been lately Visited by Persons of our own Nation, and of unquestionable Fidelity; whereby the Work is rendred very *Useful* and *Entertaining*. Illustrated and Adorned with Maps and several *Copper-Plates*; wherein is represented the *Present State* of the Places now most remarkable. By *Edward Wells*, D. D. Rector of *Cotesbach* in *Leicestershire*. Price 6 *s*.

A Demonstration of the Being and Attributes of God: more particularly in Answer to Mr *Hobbs*, *Spinoza*, and their Followers. Wherein the Notion of *Liberty* is stated, and the Possibility and Certainty of it proved, in Opposition to *Necessity* and *Fate*. Being the Substance of Eight Sermons Preach'd at the Cathedral-Church of St. *Paul*, in the Year 1704. at the Lecture Founded by the Honourable *Robert Boyle* Esq; *The Second Edition.* Price 3 *s*.

A Discourse concerning the Unchangeable Obligation of Natural Religion, and the Truth and Certainty of the *Christian Revelation.* Being Eight Sermons Preach'd at the Cathedral-Church of St *Paul*, in the Year 1705, at the Lecture Founded by the Honourable *Robert Boyle* Esq; The Second Edition. Price 5 *s*.

A Paraphrase on the Four *Evangelists*. Wherein, for the clearer Understanding the Sacred History, the whole Text and Paraphrase are Printed in separate Columes over-against each other. Together with Critical Notes on the more

more difficult Passages. Very useful for Families. In two Volumes. 8*vo*. Price 12 *s*.

The whole Duty of a Christian, Plainly represented in three Practical Essays, on *Baptism*, *Confirmation* and *Repentance*. Containing full Instructions for a Holy Life: With earnest Exhortations, especially to young Persons, drawn from the Consideration of the Severity of the Discipline of the Primitive Church. The Second Edition. Price 6 *d*. 100 for 2 *l*. fine Paper Bound 1 *s*.

Some Reflexions on that part of a Book called *Amyntor*, or, The Defense of *Miltons* Life, which relates to the Writings of the Primitive Fathers and the Canon of the New Testament. *In a Letter to a Friend*. Octavo. Pr. 6 *d*.

The Great Duty of Universal Love and Charity. A Sermon Preached before the Queen, at St *James*'s Chapel. On *Sunday December* the 30th, 1705 Price 6 *d*.

A Letter to Mr *Dodwel*, &c. The third Edition. Pr. 1 *s*.

A Defense of it, &c. Price 6 *d*.

——— 2*d* Defense of it. Price 6*d*.

——— 3*d* Defense. Price 1 *s*.

——— 4*th* Defense. Price 1 *s*.

Jacobi Rohaulti Physica. Latine vertit, recensuit, & uberioribus jam Annotationibus ex illustrissimi *Isaaci Newtoni* Philosophia maximam partem haustis, amplificavit & ornavit *S*, *C*. Accedunt etiam in hac secunda Editione, novæ aliquot Tabulæ æri incisæ. 8*vo*. Price 8 *s*.

All these by the Reverend Mr. Clark.

Devotions *viz*. Confessions, Petitions, Intercessions, and Thanksgivings for every Day of the Week; and also Before, At, and After the Sacrament: With Occasional Prayers for all Persons whatsoever. By *Thomas Bennet*, M. A. Rector of St. *James*'s in *Colchester*, and Fellow of St. *John*'s College in *Cambridge*.

Bennet of Schism. Price 2 *s*. 6 *d*.

——Defence of it. pr. 1 *s*.

Confutation of Popery. pr. 4 *s*.

———Of Quakerism. pr. 4 *s*.

History of Prayer. pr. 5 *s*.

——On Joint Prayer. pr. 2 *s*. 6 *d*.

——His Paraphrase on the Common-Prayer. pr. 4 *s*.

Dampier's

DAMPIER's Voyages.

VOL. III.

PART II.

CHAP. I.

The A's departure from the Coast of New Holland, *with the Reasons of it. Water-Snakes. The A's arrival at the Island* Timor. *Search for fresh Water on the South-side of the Island, in vain. Fault of the Charts. The Island* Rotee. *A Passage between the Islands* Timor *and* Anabao.

An. 1699.

Fault of the Charts. A Dutch Fort, called Concordia. *Their Suspicion of the A. The Island* Anabao *described. The A's Parly with the Governour of the Dutch Fort. They, with great difficulty, obtain leave to Water.* Copang *Bay. Coasting along the North side of* Timor. *They find Water and an Anchoring-place. A Description of a small Island, seven Leagues East from the Watering Bay.* Laphao *Bay. How the A. was treated by the* Portugueze *there. Designs of making further searches upon and about the Island. Port* Sesial. *Return to* Babao *in* Copang *Bay. The* A's *entertainment at the Fort of* Concordia. *His stay seven weeks at* Babao.

I Had spent about five Weeks in ranging off and on the Coast of *New Holland*, a Length of about three hundred Leagues; and had put in at three several places, to see what there might be

be thereabouts worth discovering; and
at the same time to recruit my stock of fresh Water and Provisions for the further Discoveries I purposed to attempt on the *Terra Australis*. This large and hitherto almost unkown Tract of Land, is situated so very advantageously in the richest Climates of the World, the *Torrid* and *Temperate Zones*; having in it especially all the advantages of the *Torrid Zone*, as being known to reach from the *Equator* it self (within a Degree) to the *Tropick* of *Capricorn*, and beyond it; that in coasting round it, which I design'd by this Voyage, if possible; I could not but hope to meet with some fruitful Lands, Continent or Islands, or both, productive of any of the rich Fruits, Drugs, or Spices, (perhaps Minerals also, *&c.*) that are in the other parts of the *Torrid Zone*, under equal Parallels of Latitude; at least a Soil and Air capable of such, upon transplanting them hither, and Cultivation. I meant also to make as diligent a Survey as I could, of the several smaller Islands, Shores, Capes, Bays, Creeks, and Harbours, fit as well for Shelter as Defense, upon fortifying them; and of the Rocks and Sholes, the Soundings, Tides, and Currents, Winds and Weather, Variation, *&c.* Whatever might be bene-

An. 1699. ficial for Navigation, Trade, or Settlement; or be of uſe to any who ſhould proſecute the ſame Deſigns hereafter; to whom it might be ſerviceable to have ſo much of their work done to their hands; which they might advance and perfect by their own repeated Experiences; as there is no Work of this kind brought to perfection at once. I intended eſpecially to obſerve what Inhabitants I ſhould meet with, and to try to win them over to ſomewhat of Traffick and uſeful Intercourſe, as there might be Commodities among any of them that might be fit for Trade or Manufacture, or any found out in which they might be employed. Though as to the *New Hollanders* hereabouts, by the Experience I had had of their Neighbours formerly, I expected no great matters from them.

With ſuch Views as theſe, I ſet out at firſt from *England*; and would, according to the Method I propoſed formerly [Vol. I.] have gone Weſtward, through the *Magellanick* Streight, or round *Terra del Fuego* rather, that I might have begun my Diſcoveries upon the Eaſtern and leaſt known ſide of the *Terra Auſtralis.* But that way 'twas not poſſible for me to go, by reaſon of the time of Year in which I came out: For I muſt have been compaſſing the South of *America* in a very

very high Latitude, in the depth of the Winter there. I was therefore necessitated to go Eastward by the *Cape* of *Good Hope*; and when I should be past it, 'twas requisite I should keep in a pretty high Latitude, to avoid the general Trade-winds that would be against me, and to have the benefit of the Variable Winds: By all which I was in a manner unavoidably determin'd to fall in first with those parts of *New Holland* I have hitherto been describing. For should it be ask'd why at my first making that Shore, I did not coast it to the Southward, and that way try to get round to the East of *New Holland* and *New Guinea*; I confess I was not for spending my time more than was necessary in the higher Latitudes; as knowing that the Land there could not be so well worth the discovering, as the Parts that lay nearer the Line, and more directly under the Sun. Besides, at the time when I should come first on *New Holland*, which was early in the Spring, I must, had I stood Southward, have had for some time a great deal of Winter-weather, increasing in severity, though not in time, and in a place altogether unknown; which my Men, who were heartless enough to the Voyage at best, would never have born, after

An. 1699. ſo long a Run as from *Brazil* hither.

For theſe Reaſons therefore I choſe to coaſt along to the Northward, and ſo to the Eaſt, and ſo thought to come round by the South of *Terra Auſtralis* in my return back, which ſhould be in the Summer-ſeaſon there; And this Paſſage back alſo I now thought I might poſſibly be able to ſhorten, ſhould it appear, at my getting to the Eaſt Coaſt of *New Guinea*, that there is a Channel there coming out into theſe Seas, as I now ſuſpected, near *Roſemary Iſland*: Unleſs the high Tides and great Indraught thereabout ſhould be occaſion'd by the Mouth of ſome large River; which hath often low Lands on each ſide of its Outlet, and many Iſlands and Sholes lying at its Entrance. But I rather thought it a Channel or Streight, than a River: And I was afterwards confirmed in this Opinion, when, by coaſting *New Guinea*, I found that other parts of this great Tract of *Terra Auſtralis*, which had hitherto been repreſented as the Shore of a Continent, were certainly Iſlands; and 'tis probably the ſame with *New Holland*: Though for Reaſons I ſhall afterwards ſhew, I could not return by the way I propos'd to my ſelf, to fix the Diſcovery. All that I had now ſeen from the

La-

Latitude of 27 d. South to 25, which is *Sharks-Bay*; and again from thence to *Rosemary Islands*, and about the Latitude of 20; seems to be nothing but Ranges of pretty large Islands against the Sea, whatever might be behind them to the Eastward, whether Sea or Land, Continent or Islands.

But to proceed with my Voyage. Though the Land I had seen as yet, was not very inviting, being but barren towards the Sea, and affording me neither fresh Water, nor any great store of other Refreshments, nor so much as a fit place for careening; yet I stood out to Sea again, with thoughts of coasting still along Shore (as near as I could) to the North Eastward, for the further discovery of it: Perswading my self, that at least the place I anchor'd at in my *Voyage round the World*, in the Latitude of 16 deg. 15 min. from which I was not now far distant, would not fail to afford me sweet Water upon digging, as it did then; For the brackish Water I had taken in here, though it serv'd tolerably well for boiling, was yet not very wholsome.

With these Intentions I put to Sea on the 5th of *September* 1699, with a gentle Gale, sounding all the way; but was quickly induc'd to alter my design. For

An. 1699. I had not been out above Day, but I found that the Sholes, among which I was engaged all the while on the Coast, and was like to be engag'd in, would make it a very tedious thing to sail along by the Shore, or to put in where I might have occasion. I therefore edged farther off to Sea, and so deepned the Water from eleven to thirty-two Fathom. The next day, being *September* the 6th, we could but just discern the Land, though we had then no more than about thirty Fathom, uncertain Soundings; For even while we were out of sight of Land, we had once but seven Fathom, and had also great and uncertain Tides whirling about, that made me afraid to go near a Coast so shallow, where we might be soon a-ground, and yet have but little Wind to bring us off: For should a Ship be near a Shole, she might be hurl'd upon it unavoidably by a strong Tide, unless there should be a good Wind to work her and keep her off. Thus also on the seventh day we saw no Land, though our Water decreas'd again to twenty-six Fathom; for we had deepned it, as I said, to thirty.

This Day we saw two Water-snakes, different in shape from such as we had formerly seen. The one was very small, though long; the other long and as big

as

as a Mans Leg, having a red Head; which I never saw any have, before or since. We had this Day, Lat. 16 d. 9 m. by Observation.

An. 1699.

I was by this time got to the North of the Place I had thought to have put in at, where I dug Wells in my former Voyage; and though I knew by the Experience I had of it then, that there was a deep entrance in thither from the Eastward; yet by the Sholes I had hitherto found so far stretcht on this Coast, I was afraid I should have the same Trouble to coast all along afterwards beyond that place: And besides the danger of running almost continually amongst Sholes on a strange Shore, and where the Tides were strong and high; I began to bethink my self, that a great part of my Time must have been spent in beating about a Shore I was already almost weary off, which I might employ with greater satisfaction to my mind, and better hopes of success, in going forward to *New Guinea*. Add to this the particular danger I should have been in upon a Lee-Shore, such as is here describ'd, when the North-West Monsoon should once come in; the ordinary season of which was not now far off, though this Year it staid beyond the common season: And it comes on storming at first, with Tor-

nadoes,

An. 1699. nadoes, violent Gusts, &c. Wherefore quitting the thoughts of putting in again at *New Holland*, I resolv'd to steer away for the Island *Timor*; where, besides getting fresh Water, I might probably expect to be furnished with Fruits, and other Refreshments to recruit my Men, who began to droop; some of them being already to my great grief, afflicted with the Scurvy, which was likely to increase upon them and disable them, and was promoted by the Brackish Water they took in last for boiling their Oatmeal. 'Twas now also towards the latter end of the dry season; when I might not probably have found Water so plentifully upon Digging at that part of *New Holland*, as when I was there before in the wet season. And then, considering the time also that I must necessarily spend in getting in to the Shore, through such Sholes as I expected to meet with; or in going about to avoid them; and in digging of Wells when I should come thither: I might very well hope to get to *Timor*, and find fresh Water there, as soon as I could expect to get it at *New Holland*; and with less trouble and danger.

On the 8th of *September* therefore, shaping our Course for *Timor*, we were in Lat. 15 d. 37 m. We had twenty six fathom,

fathom, Course-sand; and we saw one Whale. We found them lying most commonly near the Shore, or in Shole Water. This day we also saw some small white Clouds; the first that we had seen since we came out of *Sharks* Bay. This was one sign of the approach of the North-North-West Monsoon. Another sign was the shifting of the Winds; for from the time of our coming to our last Anchoring place, the Sea-Breezes which before were Easterly and very strong, had been whiffling about and changing gradually from the East to the North, and thence to the West, blowing but faintly, and now hanging mostly in some point of the West. This Day the Winds were at South-West by West, blowing very faint; and the 9th day we had the Wind at North-West by North, but then pretty fresh; and we saw the Clouds rising more and thicker in the North West. This night at twelve we lay by for a small low sandy Island, which I reckoned my self not far from. The next morning at Sun-rising we saw it from the Top-mast-head, right a-head of us; and at noon were up within a Mile of it: When, by a good Observation, I found it to lye in 13d. 55m. I have mentioned it in my first Vol. pag. 461. but my Account then made it to lie

An. 1699.

An. 1699. lye in 13 d. 50 m. We had abundance of Boobies and *Man of War* Birds flying about us all the Day; especially when we came near the Island; which had also abundance of them upon it; though it was but a little spot of Sand, scarce a Mile round.

I did not anchor here, nor send my Boat ashore; there being no appearance of getting any thing on that spot of Sand, besides Birds that were good for little: Though had I not been in haste, I would have taken some of them. So I made the best of my way to *Timor*; and on the 11th in the afternoon we saw ten small Land-birds, about the bigness of Larks, that flew away North West. The 13th we saw a great many Sea-snakes. One of these, of which I saw great Numbers and Variety in this Voyage, was large, and all black: I never saw such another for his Colour.

We had now had for some days small Gales, from the South South West to the North North West, and the Sky still more cloudy, especially in the Mornings and Evenings. The 14th it look'd very black in the North West all the day; and a little before Sun-set we saw, to our great Joy, the tops of the high Mountains of *Timor*, peeping out of the

Clouds

Clouds, which had before covered them, as they did ſtill the lower parts. An. 1699.

We were now running directly towards the middle of the Iſland, on the South ſide: But I was in ſome doubt whether I ſhould run down along Shore on this South-ſide towards the Eaſt-end; or paſs about the Weſt-end, and ſo range along on the North-ſide, and go that way towards the Eaſt-end: But as the Winds were now Weſterly, I thought it beſt to keep on the South-ſide, till I ſhould ſee how the Weather would prove; For, as the Iſland lies, if the Weſterly Winds continued and grew tempeſtuous, I ſhould be under the Lee of it, and have ſmooth Water, and ſo could go along ſhore more ſafely and eaſily on this South-ſide: I could ſooner alſo run to the Eaſt-end, where there is the beſt ſhelter, as being ſtill more under the Lee of the Iſland when thoſe Winds blow. Or if, on the other ſide, the Winds ſhould come about again to the Eaſtward, I could but turn back again, (as I did afterwards;) and paſſing about the Weſt-end, could there proſecute my ſearch on the North ſide of the Iſland for Water, or Inhabitants, or a good Harbour, or whatever might be uſeful to me. For both ſides of the Iſland were hitherto alike to me, being wholly unacquainted

An. 1699. acquainted here; only as I had seen it at a distance in my former Voyage. [*See* Vol. I. pag. 460.]

I had heard also, that there were both *Dutch* and *Portugueze* Settlements on this Island; but whereabouts, I knew not: However, I was resolved to search about till I found, either one of these Settlements, or Water in some other place.

It was now almost Night, and I did not care to run near the Land in the dark, but clapt on a Wind, and stood off and on till the next Morning, being *September* 15th, when I steered in for the Island, which now appear'd very plain, being high, double and treble Land, very remarkable, on whatever side you view it. *See a sight of it in two parts, Table V. N°. 1. aa.* At three in the Afternoon we anchored in fourteen fathom, soft black oasy ground, about a Mile from the Shore. *See two sights more of the Coast, in Table V. N°. 2. 3. and the Island it self* in the *Particular Map*; which I have here inserted, to shew the Course of the Voyage from hence to the Eastward; as the *General Map*, set before the Title *Vol. III. Par. I.* shews the Course of the whole Voyage. But in making the *Particular Map*, I chose to begin only with *Timor*, that I might not, by extending it too far, be

forced

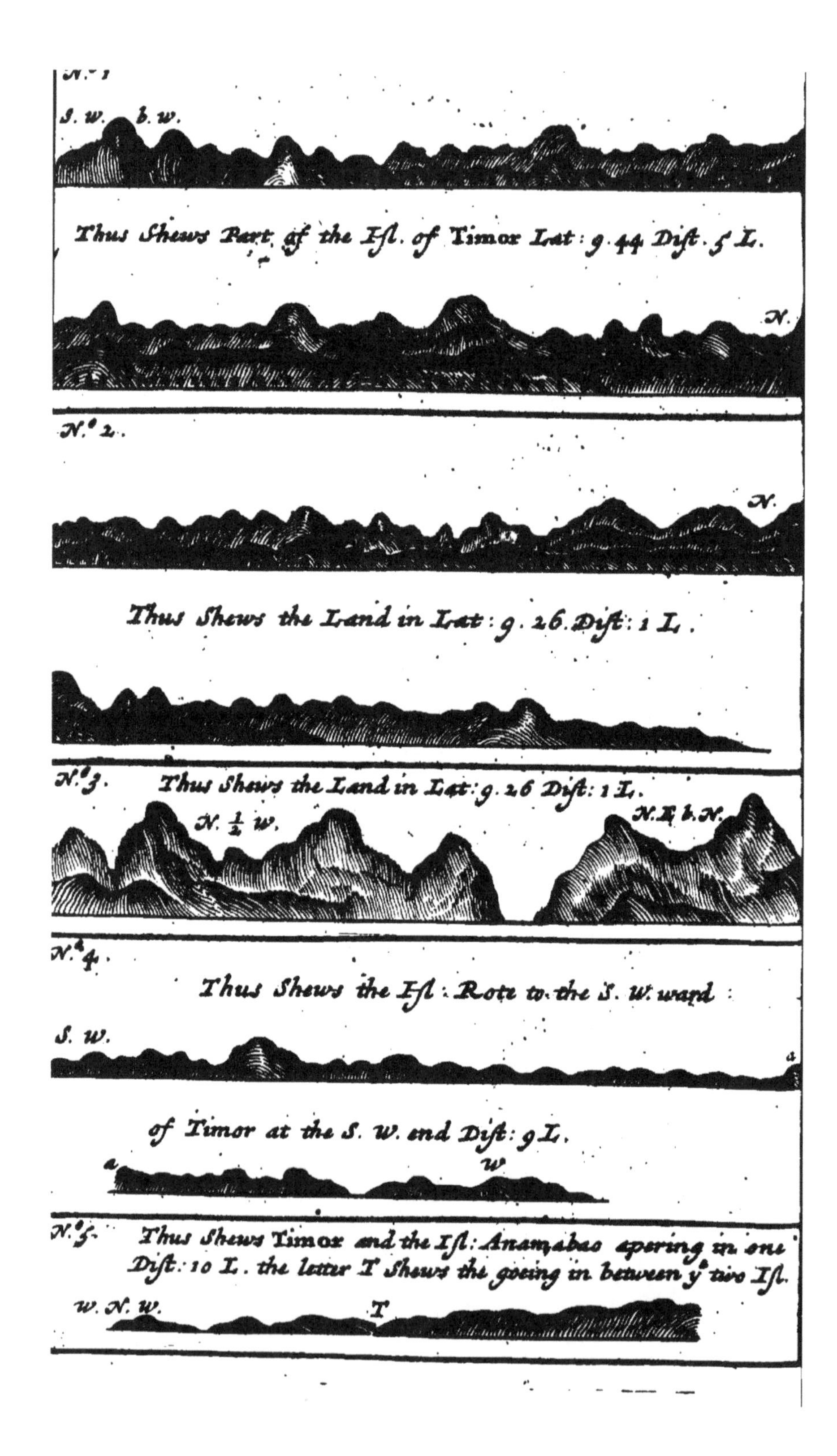
N.° 1
S. W. b. W.
Thus Shews Part of the Isl. of Timor Lat: 9. 44. Dist. 5 L.
N.
N.° 2.
N.
Thus Shews the Land in Lat: 9. 26. Dist: 1 L.
N.° 3. Thus Shews the Land in Lat: 9. 26 Dist: 1 L.
N. ½ W.
N.E. b. N.
N.° 4.
Thus Shews the Isl: Rote to the S. W. ward
S. W.
a
of Timor at the S. W. end Dist: 9 L.
a
W
N.° 5. Thus Shews Timor and the Isl: Anamabao apering in one
Dist: 10 L. the letter T Shews the goeing in between yᵉ two Isl.
W. N. W.
T

forced to contract the Scale too much a- mong the Islands, *&c.* of the *New Guinea* Coast; which I chiefly designed it for.

The Land by the Sea, on this South side, is low and sandy, and full of tall Streight-bodied Trees like Pines, for about two hundred Yards inwards from the Shore. Beyond that, further in towards the Mountains, for a Breadth of about three Miles more or less, there is a Tract of swampy Mangrovy Land, which runs all along between the sandy Land of the Shore on one side of it, and the Feet of the Mountains on the other. And this low Mangrovy Land is overflown every Tide of Flood, by the Water that flows into it through several Mouths or Openings in the outer sandy Skirt against the Sea. We came to an Anchor right against one of these Openings; and presently I went in my Boat to search for fresh Water, or get speech of the Natives; for we saw Smoaks, Houses, and Plantations against the sides of the Mountains, not far from us. It was ebbing Water before we got ashore, though the Water was still high enough to float us in without any great Trouble. After we were within the Mouth, we found a large Salt-Water Lake, which we hoped might bring us up through the

An. 1699. the Mangroves to the fast Land: But before we went further, I went ashore on the sandy Land by the Sea side, and look'd about me; but saw there no sign of fresh Water. Within the sandy Bank, the Water forms a large Lake: Going therefore into the Boat again, we rowed up the Lake towards the firm Land, where no doubt there was fresh Water, could we come at it. We found many Branches of the Lake entring within the Mangrove Land, but not beyond it. Of these we left some on the Right-hand, and some on the Left, still keeping in the biggest Channel; which still grew smaller, and at last so narrow, that we could go no farther, ending among the Swamps and Mangroves. We were then within a Mile of some Houses of the *Indian* Inhabitants, and the firm Land by the sides of the Hills; But the Mangroves thus stopping our way, we return'd as we came: But it was almost dark before we reach'd the Mouth of the Creek. 'Twas with much ado that we got out of it again; for it was now low Water, and there went a rough short Sea on the Bar; which, however, we past over without any damage, and went aboard.

The next Morning at five we weighed, and stood along Shore to the

Eastward,

Eaſtward, making uſe of the Sea and Land-Breezes. We found the Sea-Breezes here from the S. S. E. to the S. S. W. the Land-Breezes from the N. to the N. E. We coaſted along about twenty Leagues, and found it all a ſtreight, bold, even Shore, without Points, Creeks or Inlets for a Ship: And there is no anchoring till within a Mile or a Mile and an half of the Shore. We ſaw ſcarce any Opening fit for our Boats; and the faſt Land was ſtill barricado'd with Mangroves: So that here was no hope to get Water; nor was it likely that there ſhould be hereabouts any *European* Settlement, ſince there was no ſign of a Harbour.

The Land appear'd pleaſant enough to the Eye: For the ſides and tops of the Mountains were cloath'd with Woods mix'd with Savannahs; and there was a Plantation of the *Indian* Natives, where we ſaw the Coco-Nuts growing, and could have been glad to have come at ſome of them. In the Draught I had with me, a Shole was laid down hereabouts; but I ſaw nothing of it, going or coming; and ſo have taken no notice of it in my Map.

Weary of running thus fruitleſly along the South ſide of the Iſland to the Eaſtward, I reſolv'd to return the way I came; and compaſſing the Weſt end of the I-

An. 1699. ſland, make a ſearch along the North ſide of it. The rather, becauſe the North-North-Weſt Monſoon, which I had deſign'd to be ſhelter'd from by coming the way I did, did not ſeem to be near at hand, as the ordinary Seaſon of them required; but on the contrary I found the Winds returning again to the South-Eaſtward; and the Weather was fair, and ſeem'd likely to hold ſo; and conſequently the North-North-Weſt Monſoon was not like to come in yet. I conſidered therefore that by going to the North ſide of the Iſland, I ſhould there have the ſmooth Water, as being the Lee-ſide as the Winds now were; and hoped to have better riding at Anchor or Landing on that ſide, than I could expect here, where the Shore was ſo lined with Mangroves.

Accordingly, the 18th about Noon I altered my Courſe, and ſteered back again towards the South-Weſt end of the Iſland. This day we ſtruck a Dolphin; and the next day ſaw two more, but ſtruck none: We alſo ſaw a Whale.

In the Evening we ſaw the Iſland *Rotee*, and another Iſland to the South of it, not ſeen in my Map; both lying near the South-Weſt end of *Timor*. On both theſe Iſlands we ſaw Smoaks by Day,

Day, and Fires by Night, as we had seen on *Timor* ever since we fell in with it. I was told afterwards by the *Portugueze*, that they had Sugar-works on the Island *Rotee*; but I knew nothing of that now; and the Coast appearing generally dry and barren, only here and there a Spot of Trees, I did not attempt Anchoring there, but stood over again to the *Timor* Coast.

Au. 1699.

September the 21st, in the Morning, being near *Timor*, I saw a pretty large Opening, which immediately I entred with my Ship, sounding as I went in: But had no ground till I came within the East point of the Mouth of the Opening, where I Anchored in nine Fathom, a League from the Shore. The distance from the East side to the West side of this Opening, was about five Leagues. But whereas I thought this was only an Inlet or large Sound that ran a great way into the Island *Timor*, I found afterwards that it was a Passage between the West end of *Timor* and another small Island called *Anamabao* or *Anabao*: Into which Mistake I was led by my Sea-Chart, which represented both sides of the Opening as parts of the same Coast, and called all of it *Timor*: *See all this rectified, and a View of the whole Passage,*

 as

An. 1699. *as I found it, in a small Map I have made of it. Table VI. N°. 1.*

I designed to Sail into this Opening till I should come to firm Land; for the Shore was all set thick with Mangroves here by the Sea, on each side; which were very green, as were also other Trees more within Land. We had now but little Wind; therefore I sent my Boat away, to sound, and to let me know by signs what depth of Water they met with, if under eight Fathom; but if more, I order'd them to go on, and make no signs. At eleven that Morning, having a pretty fresh Gale, I weighed, and made Sail after my Boat; but edg'd over more to the West shore, because I saw many smaller Openings there, and was in hopes to find a good Harbour where I might secure the Ship: For then I could with more safety send my Boats to seek for fresh Water. I had not sailed far, before the Wind came to the South-East and blew so strong, that I could not with safety venture nearer that side, it being a Lee-shore. Besides, my Boat was on the East side of the *Timor* Coast; for the other was, as I found afterwards, the *Anabao* Shore; and the great Opening I was now in, was the Streight between that Island and *Timor*; towards which I now tack'd and stood over.

Taking

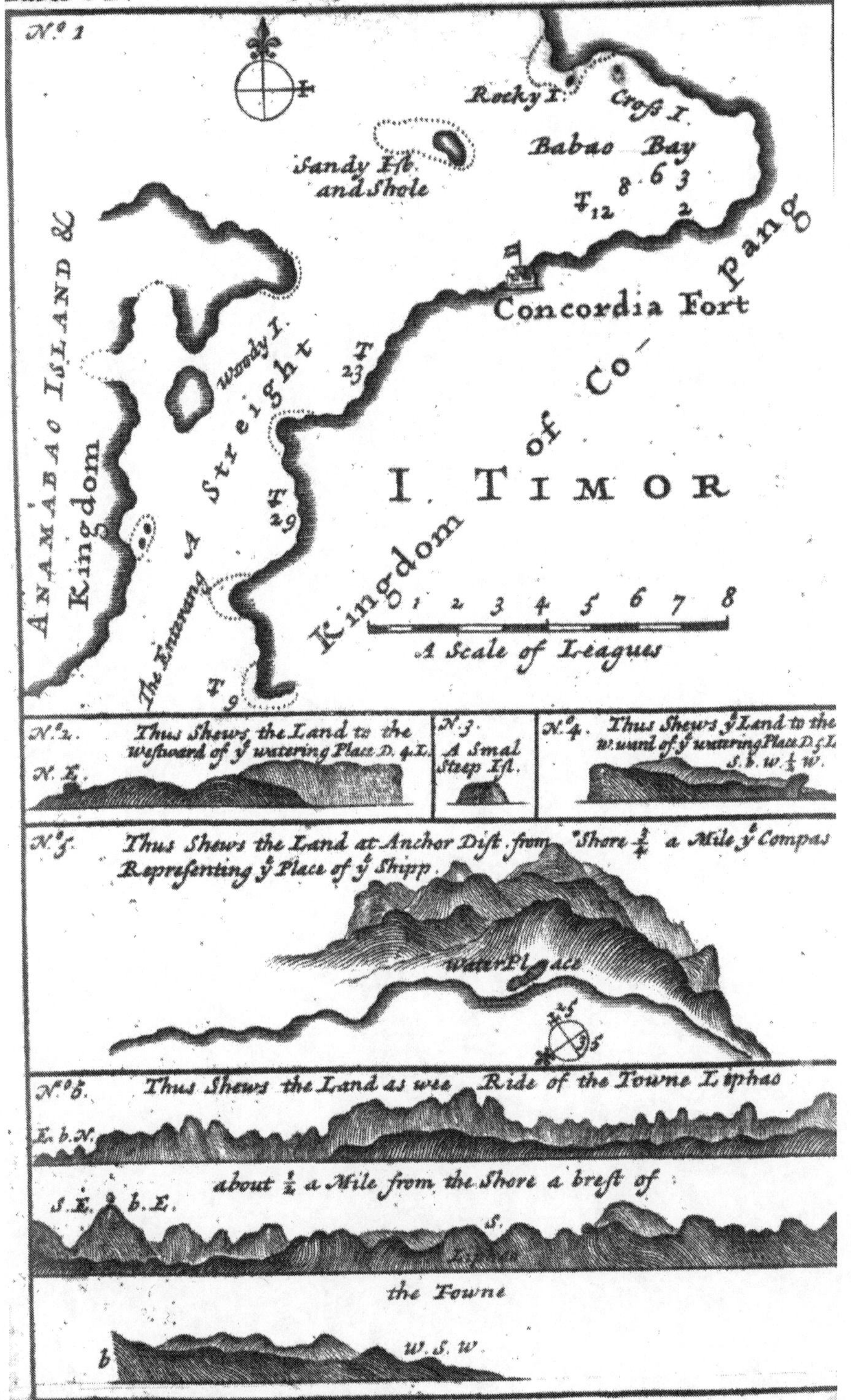
N.º 1
Rocky I.
Cross I.
Babao Bay
Sandy Isl. and Shole
Concordia Fort
ANAMABAO ISLAND & Kingdom
Woody I.
A Streight
The Entrance
I. TIMOR
Kingdom of Copang
1 2 3 4 5 6 7 8
A Scale of Leagues
N.º 2. Thus Shews the Land to the Westward of ye watering Place D. 4 L.
N. E.
N.º 3. A Smal Steep Isl.
N.º 4. Thus Shews ye Land to the W.ward of ye watering Place D. 5 L.
S. b. W. ½ W.
N.º 5. Thus Shews the Land at Anchor Dist. from ye Shore ¾ a Mile ye Compas Representing ye Place of ye Shipp.
Water Place
N.º 6. Thus Shews the Land as wee Ride of the Towne Liphao
E. b. N.
about ½ a Mile from the Shore a brest of
S. E. b. E.
S.
the Towne
W. S. W.

Taking up my Boat therefore, I ran under the *Timor* side, and at three a Clock anchored in twenty-nine Fathom, half a Mile from the Shore. That part of the South-West Point of *Timor*, where we Anchored in the Morning, bore now South by West, distance three Leagues: And another Point of the Island bore North-North-East, distance two Leagues.

An. 1699.

Not long after, we saw a Sloop coming about the Point last mention'd, with *Dutch* Colours; which I found, upon sending my Boat aboard, belonged to a *Dutch* Fort, (the only one they have in *Timor*) about 5 Leagues from hence, call'd *Concordia*. The Governour of the Fort was in the Sloop, and about forty Soldiers with him. He appear'd to be somewhat surprised at our coming this way; which it seems is a Passage scarce known to any but themselves; as he told the Men I sent to him in my Boat. Neither did he seem willing that we should come near their Fort for Water. He said also, that he did not know of any Water on all that part of the Island, but only at the Fort; and that the Natives would kill us, if they met us ashore. By the small Arms my Men carried with them in the Boat, they took us to be Pirates, and would not easily believe the Account my Men gave them of

An. 1699. of what we were, and whence we came. They said that about two Years before this, there had been a stout Ship of *French* Pirates here; and that after having been suffered to Water, and to refresh themselves, and been kindly used, they had on a sudden gone among the *Indians*, Subjects of the Fort, and plunder'd them and burnt their Houses. And the *Portugueze* here told us afterwards, that those Pirates, whom they also had entertain'd, had burnt their Houses, and had taken the *Dutch* Fort, (though the *Dutch* car'd not to own so much,) and had driven the Governour and Factory among the wild *Indians* their Enemies. The *Dutch* told my Men further, that they could not but think we had of several Nations (as is usual with Pirate Vessels) in our Ship, and particularly some *Dutch* Men, though all the Discourse was in *French*; (for I had not one who could speak *Dutch*:) Or else, since the common Draughts make no Passage between *Timor* and *Anabao*, but lay down both as one Island; they said they suspected we had plundered some *Dutch* Ship of their particular Draughts, which they are forbid to part with.

With these Jealousies the Sloop returned towards their Fort, and my Boat came back with this News to me: But I was not discouraged at this News; not doubting

doubting but I should perswade them better, when I should come to talk with them. So the next Morning I weighed, and stood towards the Fort. The Winds were somewhat against us, so that we could not go very fast, being obliged to tack two or three times: And coming near the farther end of the Passage between *Timor* and *Anabao*, we saw many Houses on each side not far from the Sea, and several Boats lying by the Shore. The Land on both sides was pretty high, appearing very dry and of a reddish Colour, but highest on the *Timor* side. The Trees on either side were but small, the Woods thin, and in many places the Trees were dry and withered.

An. 1699.

The Island *Anamabao* or *Anabao*, is not very big, not exceeding ten Leagues in length, and four in breadth; yet it has two Kingdoms in it, *viz.* that of *Anamabao* on the East-side towards *Timor*, and the North-East end; and that of *Anabao*, which contains the South-West end and the West side of the Island: but I know not which of them is biggest. The Natives of both are of the *Indian* kind, of a swarthy Copper colour, with black lank Hair. Those of *Anamabao* are in League with the *Dutch*, as these afterwards told me, and with the Natives of the Kingdom of *Copang* in *Timor*,

An. 1699. over-againſt them, in which the *Dutch* Fort *Concordia* ſtands: But they are ſaid to be inveterate Enemies to their Neighbours of *Anabao*. Thoſe of *Anabao*, beſides managing their ſmall Plantations of Roots and a few Coco-nuts, do fiſh, ſtrike Turtle, and hunt Buffalo's; killing them with Swords, Darts, or Lances. But I know not how they get their Iron; I ſuppoſe, by Traffick with the *Dutch* or *Portugueſe*, who ſend now and then a Sloop and trade thither, but well-arm'd; for the Natives would kill them, could they ſurprize them. They go always armed themſelves: And when they go a fiſhing or a hunting, they ſpend four or five Days or more in ranging about, before they return to their Habitation. We often ſaw them, after this, at theſe Employments: but they would not come near us. The Fiſh or Fleſh that they take, beſides what ſerves for preſent ſpending, they dry on a Barbacue or wooden Grate, ſtanding pretty high over the Fire, and ſo carry it home when they return. We came ſometimes afterwards to the places where they had Meat thus a drying, but did not touch any of it.

But to proceed; I did not think to ſtop any where till I came near the Fort; which yet I did not ſee: But co-

coming to the end of this Paſſage, I found that if I went any farther I ſhould be open again to the Sea. I therefore ſtood in cloſe to the Shore on the Eaſt ſide, and Anchored in four Fathom Water, ſandy ground; a point of Land ſtill hindring me from ſeeing the Fort. But I ſent my Boat to look about for it: and in a ſhort time ſhe returned, and my Men told me they ſaw the Fort, but did not go near it; and that it was not above four or five Miles from hence: It being now late, I would not ſend my Boat thither till the next Morning: Mean while about two or three hundred *Indians*, Neighbours of the Fort, and ſent probably from thence, came to the ſandy Bay juſt againſt the Ship; where thy ſtaid all Night, and made good Fires. They were armed with Lances, Swords and Targets, and made a great Noiſe all the Night: We thought it was to ſcare us from landing, ſhould we attempt it: But we took little notice of them.

The next Morning, being *September* the 23d, I ſent my Clerk aſhore in my Pinace to the Governour, to ſatisfy him that we were *Engliſh* Men, and in the *King*'s Ship, and to ask Water of him; ſending a young Man with him, who ſpake *French*. My Clerk was with the Governour pretty early; and in anſwer

to

An. 1699. to his Queries about me, and my business in these Parts, told him that I had the King of *England*'s Commission, and desired to speak with him. He beckned to my Clerk to come ashore; but assoon as he saw some small Arms in the Stern Sheets of the Boat, he commanded him into the Boat again, and would have him be gone. My Clerk sollicited him that he would allow him to speak with him; and at last the Governour consented that he should come ashore; and sent his Lieutenant and three Merchants, with a Guard of about a hundred of the Native *Indians* to receive him. My Clerk said that we were in much want of Water, and hop'd they would allow us to come to their Watering place, and fill. But the Governour replied, that he had Orders not to supply any Ships but their own *East-India Company*: neither must they allow any *Europeans* to come the way that we came; and wondred how we durst come near their Fort. My Clerk answered him, that had we been Enemies, we must have come ashore among them for Water: But, said the Governour, you are come to inspect into our Trade and Strength; and I will have you therefore be gone with all speed. My Clerk answered him, that I had no such design, but, without coming nearer

them,

them, would be contented if the Governour would ſend Water on Board where we lay, about two Leagues from the Fort; and that I would make any reaſonable ſatisfaction for it. The Governour ſaid that we ſhould have what Water we wanted, provided we came no nearer with the Ship: And ordered, that aſſoon as we pleaſed, we ſhould ſend our Boat full of empty Casks, and come to an Anchor with it off the Fort, till he ſent Slaves to bring the Casks aſhore, and fill them; for that none of our Men muſt come aſhore. The ſame Afternoon I ſent up my Boat as he had directed, with an Officer, and a Preſent of ſome Beer for the Governour; which he would not accept of, but ſent me off about a Tun of Water.

On the 24th in the Morning I ſent the ſame Officer again in my Boat; and about Noon the Boat returned again with the two principal Merchants of the Factory, and the Lieutenant of the Fort; for whoſe ſecurity they had kept my Officer and one of my Boats-crew as Hoſtages, confining them to the Governour's Garden all the time: For they were very ſhy of truſting any of them to go into their Fort, as my Officer ſaid: Yet afterwards they were not ſhy of our Company; and I found that my

Officer

An. 1699. Officer maliciously indeavour'd to make them shy of me. In the Even I gave the *Dutch* Officers that come aboard, the best Entertainment I could; and bestowing some Presents on them, sent them back very well pleased; and my Officer and the other Man were returned to me. Next Morning I sent my Boat ashore again with the same Officer; who brought me word from the Governour, that we must pay four *Spanish* Dollars, for every Boats-load of Water: But in this he spake falsly, as I understood afterwards from the Governour himself, and all his Officers, who protested to me that no such Price was demanded, but left me to give the Slaves what I pleased for their Labour: The Governour being already better satisfied about me, then when my Clerk spoke to him, or than that Officer I sent last would have caused him to be: For the Governour being a Civil, Gentile and Sensible Man, was offended at the Officer for his being so industrious to misrepresent me. I received from the Governour a little Lamb, very Fat; and I sent him two of the *Guinea*-hens that I brought from St *Jago*, of which there were none here.

I had now eleven Buts of Water on Board, having taken in seven here, which

which I would have paid for, but that at preſent I was afraid to ſend my Boat aſhore again: For my Officer told me, among other of his Inventions, that there were more Guns mounted in the Fort, than when we firſt came; and that he did not ſee the Gentlemen that were aboard the day before; intimating as if they were ſhy of us; and that the Governour was very rough with him; And I not knowing to the contrary at preſent, conſulted with my other Officers what was beſt to be done; for by this the Governour ſhould ſeem to deſign to quarrel with us. All my other Officers thought it natural to infer ſo much, and that it was not ſafe to ſend the Boat aſhore any more, leſt it ſhould be ſeiz'd on; but that it was beſt to go away, and ſeek more Water where we could find it. For having now (as I ſaid) eleven Buts aboard; and the Land being promiſing this way, I did not doubt finding Water in a ſhort time. But my Officer who occaſion'd theſe fears in us by his own Forgeries, was himſelf for going no further; having a mind, as far as I could perceive, to make every thing in the Voyage, to which he ſhew'd himſelf averſe, ſeem as Croſs and Diſcouraging to my Men as poſſible, that he might haſten our return; being very negligent

and

An. 1699. and backward in most Businesses I had occasion to employ him in; doing nothing well or willingly, though I did all I could to win him to it. He was also industrious to stir up the Sea-men to mutiny; telling them, among other things, that any *Dutch* Ship might lawfully take us in these Seas: But I knew better, and avoided every thing that could give just offence.

The rest of my Officers therefore being resolved to go from hence, and having bought some Fish of some *Anamabeans*, who, seeing our Ship, came purposely to sell some, passing to and fro every Day; I sail'd away on the 26th about five in the Afternoon. We pass'd along between a small low sandy Island (over against the Fort,) full of Bays and pretty high Trees; sounding as we went along; and had from twenty five to thirty five Fathom, oasy ground. *See the little Map of this Passage, Table VI. N°. 1.*

The 27th in the Morning we Anchored in the middle of the Bay, called *Copang* Bay, in twelve Fathom, soft oaze, about four Leagues above the *Dutch* Fort. Their Sloop was riding by the Fort, and in the Night Fired a Gun; but for what reason I know not; and the Governour said afterwards, 'twas the

Skippers

Skippers own doing, without his Order. Presently after we had Anchored, I went in the Pinnace to search about the Bay for Water, but found none. Then, returning a-board, I weighed, and ran down to the North-Entrance of the Bay, and at seven in the Evening Anchored again, in thirty seven Fathom, soft oaze, close by the sandy Island, and about four Leagues from the *Dutch* Fort. The 28th I sent both my Boats ashore on the sandy Island, to cut Wood; and by Noon they both came back laden. In the Afternoon I sent my Pinnace ashore on the North Coast or Point of *Copang* Bay, which is call'd *Babao*. Late in the Night they returned, and told me that they saw great Tracks of Buffalo's there, but none of the Buffalo's themselves; neither did they find any fresh Water. They also saw some green Turtle in the Sea, and one Alligator.

The 29th I went out of *Copang* Bay, designing to Coast it along Shore on the North side of *Timor* to the Eastward; as well to seek for Water, as also to acquaint my self with the Island, and to search for the *Portuguese* Settlements; which we were informed were about forty Leagues to the Eastward of this Place.

We

An. 1699. We coasted along Shore with Land and Sea-Breezes. The Land by the Shore was of a moderate height, with high and very remarkable Hills farther within the Country; their sides all spotted with Woods and Savannahs. But these on the Mountains sides appeared of a rusty Colour, not so pleasant and flourishing as those that we saw on the South side of the Island; For the Trees seemed to be small and withering; and the Grass in the Savannahs also look'd dry, as if it wanted moisture. But in the Valleys, and by the Sea side, the Trees look'd here also more green. Yet we saw no good Anchoring-place, or Opening, that gave us any incouragement to put in; till the 30th day in the Afternoon.

We were then running along Shore, at about four Leagues distance, with a moderate Sea-breeze; when we opened a pretty deep Bay, which appeared to be a good Road to anchor in. There were two large Valleys, and one smaller one, which descending from the Mountains came all into one Valley by the Sea side against this Bay, which was full of tall green Trees. I presently stood in with the Ship, till within two Leagues of the Shore; and then sent in my Pinnace commanded by my chief Mate, whose great care, Fidelity, and Diligence, I was well aſſured of; ordering him to ſeek for freſh Water; and if he found any, to ſound the Bay, and bring me word what Anchoring there was; and to make haſte aboard.

As ſoon as they were gone, I ſtood off a little, and lay by. The day was now far ſpent; and therefore it was late before they got aſhore with the Boat: ſo that they did not come aboard again that Night. Which I was much concern'd at; becauſe in the Evening, when the Sea-Breeze was done and the Weather calm, I perceived the Ship to drive back again to the Weſtward. I was not yet acquainted with the Tides here; for I had hitherto met with no ſtrong Tides about the Iſland, and ſcarce any running in a ſtream, to ſet me along Shore either way. But after this time, I had pretty much of them; and found at preſent the Flood ſet to the Eaſtward, and the Ebb to the Weſtward. The Ebb (with which I was now carried) ſets very ſtrong, and runs eight or nine Hours. The Flood runs but weak, and at moſt laſts not above four hours; and this too is perceived only near the Shore; where checking the Ebb, it ſwells the Seas, and makes the Water riſe in the Bays and Rivers eight or nine Foot. I was afterwards credibly informed by ſome *Portugueze*, that the Current runs

An. 1699. always to the Westward in the Mid-Channel between this Island and those that face it in a Range to the North of it, *viz.* *Misicomba* (or *Omba*) *Pintare*, *Laubana*, *Ende*, &c.

We were driven four Leagues back again, and took particular notice of a point of Land that looked like *Flambrough-head*, when we were either to the East or West of it; and near the shore, it appeared like an Island. Four or five Leagues to the East of this Point, is another very remarkable bluff Point, which is on the West side of the Bay that my Boat was in. *See two sights of this Land, Table VI. N°. II. III.* We could not stem the Tide, till about three a Clock in the Afternoon; when the Tide running with us, we soon got abreast of the Bay, and then saw a small Island to the Eastward of us. *See a sight of it, Table VI. N°. IV.* About six we Anchored in the bottom of the Bay, in twenty five Fathom, soft Oaze, half a Mile from the Shore.

I made many false Fires in the Night, and now and then fired a Gun, that my Boat might find me; but to no purpose. In the Morning I found my self driven again by the Tide of Ebb three or four Leagues to the Westward of the Place where I left my Boat. I had several Men

looking

looking out for her, but could not get sight of her: Besides, I continued still driving to the Westward; for we had but little Wind, and that against us. But by ten a Clock in the Morning we had the comfort of seeing the Boat; and at eleven she came aboard, bringing two Barrecoes of very good Water.

The Mate told me there was good Anchoring close by the Watering-place; but that there ran a very strong Tide, which near the Shore made several Races; so that they found much danger in getting ashore, and were afraid to come off again in the Night, because of the Riplings the Tide made.

We had now the Sea-breeze, and steered away for this Bay; but could hardly stemm the Tide, till about three in the Afternoon; when the Tide being turned with us, we went along briskly, and about six Anchored in the Bay, in twenty five Fathom, soft Oaze, half a Mile from the Shore.

The next Morning I went ashore to fill Water, and before Night sent aboard eight Tuns. We fill'd it out of a large Pond within fifty paces of the Sea. It look'd pale, but was very good, and boyled Pease well. I saw the Tract of an *Alligator* here. Not far from the Pond, we found the rudder of a *Malaian* Proe,

An. 1699. three great Jarrs in a small Shed set up against a Tree, and a Barbacue whereon there had been Fish and Flesh of Buffaloes drest, the Bones lying but a little from it.

In three Days we fill'd about twenty six Tun of Water, and then had on Board about thirty Tun in all. The two following days we spent in Fishing with the Saine, and the first Morning caught as many as served all my Ships Company: But afterwards we had not so good Success. The rest of my Men, which could be spared from the Ship, I sent out; Some with the Carpenters Mate, to cut Timber for my Boats, *&c*: These went always guarded with three or four armed Men to secure them: I shewed them what Wood was fitting to cut for our use, especially the Calabash and Maho; I shewed them also the manner of stripping the Maho-bark, and of making therewith Thread, Twine, Ropes, *&c.* Others were sent out a Fowling; who brought home Pidgeons, Parrots, Cackatoos, *&c.* I was always with one party or other, my self; especially with the Carpenters, to hasten them to get what they could, that we might be gone from hence.

Our Water being full, I sail'd from hence *October the* 6th about four in the Afternoon, designing to coast along

Shore

Shore to the Eaſtward, till I came to the *Portugueze* Settlements. By the next Morning we were driven three or four Leagues to the Weſt of the Bay; but in the Afternoon, having a faint Sea-breeze, we got again abreaſt of it. It was the 11th day at noon before we got as far as the ſmall Iſland before-mentioned, which lies about ſeven Leagues to the Eaſt of the Watering Bay: For what we gained in the Afternoon by the benefit of the Sea-breezes, we loſt again in the Evenings and Mornings, while it was calm, in the interval of the Breezes. But this day the Sea-breeze blowing freſher than ordinary, we paſt by the Iſland and run before Night about ſeven Leagues to the Eaſt of it.

An. 1699

This Iſland is not half a Mile long, and not above one hundred Yards in breadth, and look'd juſt like a Barn, when we were by it: It is pretty high, and may be ſeen from a Ship's Topmaſt-head about ten Leagues. The Top, and part of the ſides, are covered with Trees, and it is about three Leagues from *Timor*; 'tis about mid-way between the Watering place and the *Portugueze* firſt and main Settlement by the Shore.

In the Night we were again driven back toward the Iſland, three Leagues: But the 12th day, having a pretty briſk

An. 1699. Sea-breeze, we coasted along Shore; and seeing a great many Houses by the Sea, I stood in with my Ship till I was within two Miles of them, and then sent in my Boat, and lay by till it returned. I sent an Officer to command the Boat; and a *Portugueze* Seaman that I brought from *Brazil*, to speak with the Men that we saw on the Bay; there being a great many of them, both Foot and Horse. I could not tell what Officer there might be amongst them; but I ordered my Officer to tell the chief of them that we were *English*, and came hither for refreshment. As soon as the Boat came ashore, and the Inhabitants were informed who we were, they were very glad, and sent me word that I was welcom, and should have any thing that the Island afforded; and that I must run a little farther about a small point, where I should see more Houses; and that the Men would stand on the Bay, right against the place where I must Anchor. With this News the Boat immediately returned; adding withal, that the Governour lived about seven Miles up in the Country; and that the chief Person here was a Lieutenant, who desired me, as soon as the Ship was at Anchor, to send ashore one of my Officers to go to the Governour, and certifie him of our arrival. I presently made Sail towards the

the Anchoring place, and at five a Clock Anchored in *Laphao* Bay, in twenty Fathom, soft Oaze, over against the Town. A Description of which, and of the *Portugueze* Settlement there, shall be given in the following Chapter.

An. 1699.

Assoon as I came to Anchor, I sent my Boat ashore with my second Mate, to go to the Governour. The Lieutenant that lived here, had provided Horses and Guides for him, and sent four Soldiers with him for his Guard, and, while he was absent, treated my Men with Arack at his own House, where he and some others of the Townsmen shew'd them many broad thin pieces of Gold; telling them that they had plenty of that Metal, and would willingly traffick with them for any sort of *European* Commodities. About eleven a Clock my Mate returned on Board, and told me he had been in the Country, and was kindly received by the Gentleman he went to wait upon; who said we were welcom, and should have any thing the Island afforded; and that he was not himself the Governour, but only a Deputy. He asked why we did not salute their Fort when we anchored; My Mate answer'd that we saw no Colours flying, and therefore did not know there was any Fort till he came ashore and saw the Guns; and if we had known,

An. 1699. that there was a Fort, yet that we could not have given any Salute till we knew that they would anſwer it with the like number of Guns. The Deputy ſaid, it was very well; and that he had but little Powder; and therefore would gladly buy ſome of us, if we had any to ſpare: Which my Mate told him, we had not.

The 13th the Deputy ſent me aboard a Preſent of two young Buffaloes, ſix Goats, four Kids, an hundred and forty Coco-nuts, three hundred ripe Mangoes, and ſix ripe Jacks. This was all very acceptable; and all the time we lay here, we had freſh Proviſion, and plenty of Fruits; ſo that thoſe of my Men that were ſick of the Scurvy, ſoon recover'd and grew luſty. I ſtaid here till the 22d, went aſhore ſeveral times, and once purpoſely to ſee the Deputy; who came out of the Country alſo on purpoſe to ſee and talk with me. And then indeed there were Guns fired for Salutes, both aboard my Ship and at the Fort. Our Interview was in a ſmall Church, which was fill'd with the better ſort of people; the poorer ſort thronging on the outſide, and looking in upon us: For the Church had no Wall but at the Eaſt end; the Sides and the Weſt end being open, ſaving only that it had Boards about three or four Foot high from the Ground. I ſaw but

two

two White Men among them all; One was a *Padre* that came along with the Lieutenant; the other was an Inhabitant of the Town. The rest were all Copper-colour'd, with black lank Hair. I staid there about two Hours, and we spoke to each other by an Interpreter. I asked particularly about the Seasons of the Year, and when they expected the North-North-West Monsoon. The Deputy told me, that they expected the Wind to shift every Moment; and that some Years the North-North-West Monsoon set in in *September*, but never failed to come in *October*; and for that reason desir'd me to make what haste I could from hence; for that 'twas impossible to ride here when those Winds came. I asked him if there was no Harbour hereabouts, where I might be secured from the Fury of these Winds at their first coming. He told me, that the best Harbour in the Island was at a place called *Babao*, on the North side of *Copang* Bay; that there were no Inhabitants there, but plenty of Buffaloes in the Woods, and abundance of Fish in the Sea; that there was also fresh Water: That there was another place, call'd *Port Sesiall*, about twenty Leagues to the Eastward of *Laphao*; that there was a River of fresh Water there, and plenty

of

An. 1699. of Fish, but no Inhabitants: Yet that, if I would go thither, he would send people with Hogs, Goats and Buffaloes, to truck with me for such Commodities as I had to dispose of.

I was afterwards told, that on the East end of the Island *Ende* there was also a very good Harbour, and a *Portugueze* Town; that there was great plenty of Refreshments for my Men, and Dammer for my Ship; that the Governour or Chief of that place, was call'd Captain *More*; that he was a very courteous Gentleman, and would be very glad to entertain an *English* Ship there; and if I design'd to go thither, I might have Pilots here that would be willing to carry me, if I could get the Lieutenants consent. That it was dangerous going thither without a Pilot, by reason of the violent Tides that run between the Islands *Ende* and *Solor*. I was told also, that at the Island *Solor* there were a great many Dutchmen banisht from other places for certain Crimes. I was willing enough to go thither, as well to secure my Ship in a good Harbour, where I might careen her, (there being Dammer also, which I could not get here, to make use of instead of Pitch, which I now wanted,) and where I might still be refreshing my Men and supporting them, in order to my further

Disco-

Discoveries ; as also to inform my self more particularly concerning these places as yet so little knovvn to us. Accordingly I accepted the offer of a Pilot and tvvo Gentlemen of the Tovvn, to go vvith me' to *Larentucka* on the Island *Ende* : And they vvere to come on board my Ship the Night before I sailed. But I vvas hindred of this design by some of my Officers, vvho had here also been very busie in doing me all the injury they could underhand.

But to proceed. While I staid here, I vvent ashore every day, and my Men took their turns to go ashore and traffick for vvhat they had occasion for ; and were now all very well again : And to keep themselves in heart, every Man bought some Rice, more or less, to recruit them after our former Fatigues. Besides, I order'd the Purser to buy some for them, to serve them instead of Pease, which were now almost spent. I fill'd up my Water-Cask again here, and cut more Wood ; and sent a Present to the Lieutenant, *Alexis Mendosa*, designing to be gone ; for while I lay here, we had some Tornadoes and Rain, and the Sky in the North-West looked very black Mornings and Evenings, with Lightning all Night from that Quarter : Which made me very uneasie and desirous to depart hence ; because

An. 1699 because this Road lay expos'd to the North-North-West and North VVinds, which were now daily expected, and which are commonly so violent, that 'tis impossible for any Ship to ride them out: Yet, on the other hand, it was absolutely necessary for me to spend about 2 Months time longer in some place hereabouts, before I could prosecute my Voyage farther to the Eastward; for Reasons which I shall give hereafter in its proper place in the ensuing Discourse. When therefore I sent the Present to the Governour, I desired to have a Pilot to *Larentucka* on the Island *Ende*; where I desir'd to spend the time I had to spare. He novv sent me vvord that he could not vvell do it, but vvould send me a Letter to *Port Sesiall* for the Natives, vvho vvould come to me there and supply me vvith vvhat Provision they had.

I staid three days, in hopes yet to get a Pilot for *Larentucka*, or at least the Letter from the Governour to *Port Sesiall*. But seeing neither, I sail'd from hence the 22d of *October*, coasting to the Eastward, designing for *Sesiall*; and before Night, was about ten Leagues to the East of *Laphao*. I kept about three Leagues off Shore, and my Boat ranged along close by the Shore, looking into every Bay and Cove; and at Night returned on Board.

The

The next Morning, being three or four Leagues farther to the Eastward, I sent my Boat ashore again to find *Sesiall*. At noon they returned, and told me they had been at *Sesiall*, as they guess'd; that there were two *Portugueze* Barks in the Port, who threatned to Fire at them, but did not; telling them this was *Porto del Roy de Portugal*. They saw also another Bark, which ran and anchor'd close by the Shore; and the Men ran all away for fear: But our Men calling to them in *Portugueze*, they at last came to them, and told them that *Sesiall* was the place which they came from, where the two Barks lay: Had not these Men told them, they could not have known it to be a Port, it being only a little bad Cove, lying open to the North; having two ledges of Rocks at its Entrance, one on each side; and a Channel between, which was so narrow, that it would not be safe for us to go in. However I stood in with the Ship, to be better satisfied; and when I came near it, found it answer my Mens Description. I lay by a-while, to consider what I had best do; for my design was to lye in a place where I might get fresh Provisions if I could: For though my Men were again pretty well recruited; and those that had been sick of the Scurvy, were well again; yet I design'd, if possible,

An. 1699.

An. 1699. possible, to refresh them as much and as long as I could, before I went farther. Besides, my Ship wanted cleaning; and I was resolved to clean her, if possible.

At last after much consideration, I thought it safer to go away again for *Babao*; and accordingly stood to the Westward. We were now about sixty Leagues to the East of *Babao*. The Coast is bold all the way, having no Sholes, and but one Island which I saw and describ'd coming to the Eastward. The Land in the Country is very Mountainous; but there are some large Valleys towards the East end. Both the Mountains and Valleys on this side, are barren; some wholly so; and none of them appear so pleasant as the place where I watered. It was the 23d day in the Evening when I stood back again for *Babao*. We had but small Sea and Land-breezes. On the 27th we came into *Copang* Bay; and the next day having sounded *Babao* Road, I ran in and came to an Anchor there, in twenty Fathom, soft oaze, three Miles from the Shore. One reason, as I said before, of my coming hither, was to ride secure, and to clean my Ships bottom; as also to endeavour by Fishing and Hunting of Buffaloes, to refresh my Men and save my Salt Provision. It was like to be some time before I could clean my Ship, because

because I wanted a great many necessaries, especially a Vessel to careen by. I had a long Boat in a frame, that I brought out of *England*, by which I might have made a Shift to do it: But my Carpenter was uncapable to set her up. Besides, by that time the Ships sides were Calk'd, my Pitch was almost spent; which was all owing to the Carpenters wilful waste and ignorance; so that I had nothing to lay on upon the Ship's bottom. But instead of this, I intended to make Lime here, which with Oyl would have made a good Coat for her. Indeed had it been adviseable, I would have gone in between *Cross* Island and *Timor*; and have hal'd my Ship ashore; for there was a very convenient place to do it in; But my Ship being sharp, I did not dare to do it: Besides, I must have taken every thing out of her; and I had neither Boats to get my things ashore, nor hands to look after them when they were there; For my Men would have been all employed; and though here are no *Indians* living near, yet they come hither in Companies when Ships are here, on purpose to do any Mischief they can to them: And 'twas not above two Years since a *Portugueze* Ship riding here, and sending her Boat for Water to one of the Gallyes, the Men were all killed by the *Indians*. But to

secure

An. 1699. secure my Men, I never suffer'd them to go ashore unarmed; and while some were at work, others stood to guard them.

We lay in this place from *October the* 28th, till *December the* 12th. In which time we made very good Lime with Shells, of which here are plenty. We cut Palmeto-leaves to burn the Ship's sides; and giving her as good a heel as we could, we burned her sides, and paid them with Lime and Water for want of Oyl to mix with it. This stuck on about two Months, where 'twas well burned. We did not want fresh Provisions all the time we lay here, either of Fish or Flesh. For there were fair sandy Bays on the Point of *Babao*, where in 2 or 3 hours in a Morning we used with our Sain to drag ashore as much Fish as we could eat all the day: And for a change of Diet, when we were weary of Fish, I sent ten or eleven armed Men a hunting for Buffaloes; who never came empty home. They went ashore in the Evening or early in the Morning, and before Noon always returned with their burdens of *Buffalo*, enough to suffice us two days; by which time we began to long for Fish again.

On the 11th of *November*, the Governour of *Concordia* sent one of his Officers to us, to know who we were. For I had not sent thither, since I came to Anchor

last

laſt here. When the Officer came aboard, he ask'd me why we fired ſo many Guns the 4th and 5th days; (which we had done in Honour of King *William*, and in Memory of the deliverance from the Powder-Plot:) I told him the occaſion of it; and he replied that they were in ſome fear at the Fort that we had been *Portugueze*, and that we were coming with Soldiers to take their Fort: He asked me alſo why I did not ſtay and fill my Water at their Fort, before I went away from thence: I told him the reaſon of it, and withal offered him Money; bidding him take what he thought reaſonable: He took none, and ſaid he was ſorry there had been ſuch a miſunderſtanding between us; and knew that the Governour would be much concerned at it. After a ſhort ſtay, he went aſhore; and the next Morning came aboard again, and told me the Governour deſired me to come aſhore to the Fort and dine with him; and, if I doubted any thing, he would ſtay aboard till I returned. I told him I had no reaſon to miſtruſt any thing againſt me, and would go aſhore with him; ſo I took my Clerk and my Gunner, and went aſhore in my Pinnace: The Gunner ſpoke very good *French*, and therefore I took him to be my Interpreter, because the Governour ſpeaks *French*:

An. 1699. He was an honest Man, and I found him always diligent and obedient. It was pretty late in the Afternoon before we came ashore; so that we had but little time with the Governour. He seem'd to be much dissatisfied at the report my Officer had made to me; (of which I have before given an account;) and said it was false, neither would he now take any Money of me; but told me I was welcom; as indeed I found by what he provided. For there was plenty of very good Victuals, and well drest; and the Linnen was white and clean; and all the Dishes and Plates, of Silver or fine China. I did not meet any where with a better Entertainment, while I was abroad; nor with so much decency and order. Our Liquor was Wine, Beer, Toddy, or Water, which we liked best after Dinner. He shew'd me some drawers full of Shells, which were the strangest and most curious that I had ever seen. He told me, before I went away, that he could not supply me with any Naval stores; but if I wanted any fresh Provision, he would supply me with what I had occasion for. I thank'd him, and told him I would send my Boat for some Goats and Hogs, though afterwards on second thoughts I did not do it: For 'twas a great way from the place where we lay, to the Fort; and I could

not

not tell what mischief might befall any of my Men, when there, from the Natives; especially if incouraged by the *Dutch*, who are Enemies to all *Europeans* but such as are under their own Government. Therefore I chose rather to Fish and Hunt for Provisions, than to be beholden to the *Dutch*, and pay dearly for it too.

We found here, as I said before, plenty of Game; so that all the time we lay at this place, we spent none or very little of our Salt-provisions; having Fish or fresh Buffaloe every day. We lay here seven Weeks; and although the North-North-West Monsoon was every day expected when I was at *Laphao*, yet it was not come, so that if I had prosecuted my Voyage to the Eastward without staying here, it had been but to little advantage. For if I had gone out, and beaten against the Wind a whole Month, I should not have got far; it may be forty, fifty, or sixty Leagues; which was but twenty four hours run for us with a large Wind; besides the trouble and discontent, which might have arisen among my Men in beating to Windward to so little purpose, there being nothing to be got at Sea; but here we lived and did eat plentifully every day without trouble. The greatest inconveniency of this place, was want of Water; this being

An. 1699. the latter part of the dry Season, because the Monsoon was very late this Year. About four days before we came away, we had Tornadoes, with Thunder, Lightning and Rain, and much Wind; but of no long continuance: At which time we filled some Water. We saw very black Clouds, and heard it thunder every day for near a Month before, in the Mountains; and saw it rain, but none came near us: And even where we hunted, we saw great Trees torn up by the Roots, and great havock made among the Woods by the Wind; yet none touched us.

CHAP.

CHAP. II.

A particular Description of the Island Timor. *Its Coast. The Island* Anabao. *Fault of the Draughts. The Channel between* Timor *and* Anabao. Copang-*bay. Fort* Concordia. *A particular description of the Bay. The Anchoring-place, called* Babao. *The Malayans here kill all the Europeans they can.* Laphao, *a Portugueze Settlement, described. Port* Ciccale. *The Hills, Water, Low-lands, Soil, Woods, Metals, in the Island* Timor. *Its Trees.* Cana-fistula-*tree described. Wild Fig-trees described. Two new sorts of Palm-trees described. The Fruits of the Island. The Herbs. Its Land-Animals. Fowls. The Ringing Bird. Its Fish. Cockle-merchants and Oysters. Cockles as big as a Man's Head. Its original Natives described. The Portugueze*

An. 1699.

and Dutch Settlements. The Malayan Language generally ſpoken here. L' Orantua *on the Iſland* Ende. *The Seaſons, Winds, and Weather at* Timor.

THE Iſland *Timor*, as I have ſaid in my Voyage round the World, is about ſeventy Leagues long, and fourteen or ſixteen broad. It lies nearly North-Eaſt and South-Weſt. The middle of it lies in about 9d. South Lat. It has no Navigable Rivers, nor many Harbours; but abundance of Bays, for Ships to ride in at ſome Seaſons of the Year. The Shore is very bold, free from Rocks, Shoals or Iſlands; excepting a few which are viſible, and therefore eaſily avoided. On the South ſide there is a Shole laid down in our Draughts, about thirty Leagues from the South-Weſt end; I was fifteen or twenty Leagues further to the Eaſt than that diſtance, but ſaw nothing of the Shole; neither could I find any Harbour. It is a pretty even Shore, with Sandy Bays and low Land for about three or four Mile up; and then 'tis Mountainous. There is no Anchoring but within half a League or a League at fartheſt from the Shore; and the low Land that bounds the Sea, hath nothing but red Man-

Mangroves, even from the Foot of the Mountains till you come within a hundred and fifty or two hundred paces of the Sea; and then you have Sand-banks, cloath'd with a sort of Pine; so that there is no getting Water on this side, because of the Mangroves.

At the South-West end of *Timor*, is a pretty high Island, called *Anabao*. It is about ten or twelve Leagues long, and about four broad; near which the *Dutch* are settled. It lies so near *Timor*, that 'tis laid down in our Draughts as part of that Island; yet we found a narrow deep Channel fit for any Ships to pass between them. This Channel is about ten Leagues long, and in some places not above a League wide. It runs North-East and South-West, so deep that there is no Anchoring but very nigh the Shore. There is but little Tide; the Flood setting North, and the Ebb to the Southward. At the North-East end of this Channel, are two points of Land, not above a League asunder; one on the South side upon *Timor*, called *Copang*; the other on the North side, upon the Island *Anabao*. From this last point, the Land trends away Northerly two or three Leagues, opens to the Sea, and then bends in again to the Westward.

An. 1699. Being paſt theſe Points, you open a Bay of about eight Leagues long, and four wide. This Bay trends in on the South ſide North Eaſt by Eaſt from the South-point before mentioned; making many ſmall Points or little Coves. About a League to the Eaſt of the ſaid South-point, the *Dutch* have a ſmall Stone Fort, ſituated on a firm Rock cloſe by the Sea: This Fort they call *Concordia*. On the Eaſt ſide of the Fort, there is a ſmall River of freſh Water, which has a broad boarded Bridge over it, near to the entry into the Fort. Beyond this River is a ſmall ſandy Bay, where the Boats and Barks land and convey their Traffick in or out of the Fort. About an hundred Yards from the Sea-ſide, and as many from the Fort, and forty Yards from the Bridge on the Eaſt ſide, the Company have a fine Garden, ſurrounded with a good Stone-Wall; In it is plenty of all ſorts of Sallads, Cabbages, Roots for the Kitchen; in ſome parts of it are Fruit-trees, as Jaca's, Pumplenoſe, Oranges, ſweet Lemons, *&c.* and by the Walls are Coco-nut and Toddy-trees in great plenty. Beſides theſe, they have Musk and Water-Melons, Pine-Apples, Pomecitrons, Pomegranates, and other ſorts of Fruits. Between this Garden and the River, there is a Penn for black Cattle, whereof they have

have plenty. Beyond the Companies ground, the Natives have their Houses, in number about fifty or sixty. There are forty or fifty Soldiers belonging to this Fort, but I know not how many Guns they have; For I had only opportunity to see one Bastion, which had in it four Guns. Within the Walls there is a neat little Church or Chapel.

Beyond *Concordia* the Land runs about seven Leagues to the bottom of the Bay; then it is not above a League and half from side to side, and the Land trends away Northerly to the North Shore; then turns about again to the Westward, making the South side of the Bay. About three Leagues and a half from the bottom of the Bay on this side, there is a small Island about a Musket shot from the Shore; and a riff of Rocks that runs from it to the Eastward about a mile. On the West side of the Island is a Channel of three Fathom at low Water, of which depth it is also within, where Ships may haul in and carreen. West from this Island the Land rounds away in a Bite or Elbow, and at last ends in a low point of Land, which shoots forth a ledge of Rocks a mile into the Sea, which is dry at Low-Water. Just against the low point of Land, and to the West of the ledge of Rocks, is another pretty high and rocky, yet woody Island,

An. 1699. Island, about half a mile from the low point; which Island hath a ledge of corally Rocks running from it all along to the other small Island, only leaving one Channel between them. Many of these Rocks are to be seen at Low-Water, and there seldom is Water enough for a Boat to go over them till quarter-Flood or more. Within this ledge there is two or three Fathom Water, and without it no less than ten or twelve Fathom close to the Rocks. A League without this last Rocky Island, is another small low sandy Island, about four miles from the low point, three Leagues from the *Dutch*-Fort *Concordia*, and three Leagues and a half from the South-West point of the Bay. Ships that come in this way, must pass between this low Isle and the low Point, keeping near the Isle.

In this Bay there is any depth of Water from thirty to three Fathom, very good oazy holding ground. This affords the best shelter against all Winds, of any place about the Island *Timor*. But from *March* to *October*, while either the Southerly Winds or only Land and Sea-breezes hold, the *Concordia* side is best to ride in; but when the more violent Northerly Winds come, then the best riding is between the two Rocky Islands in nineteen or twenty Fathom. If you bring the

Wester-

Westermost Island to bear South-West by West about a League distance, and the low point West by South; then the Body of the sandy Island will bear South-West half West, distance two Leagues; and the ledges of Rocks shooting from each, make such a Bar, that no Sea can come in. Then you have the Land from West by South to East-North-East, to defend you on that side: And other Winds do not here blow violently. But if they did, yet you are so Land-lock'd, that there can be no Sea to hurt you. This Anchoring place is call'd *Babao*, about five Leagues from *Concordia*. The greatest inconveniency in it, is the multitude of Worms. Here is fresh Water enough to be had in the wet Season; every little Gull discharging fresh Water into the Sea. In the dry Season you must search for it in standing Ponds or Gulls; where the wild Buffaloes, Hogs, &c. resort every Morning and Evening to drink; where you may lye and shoot them, taking care that you go strong enough and well-armed against the Natives upon all occasions. For though there are no Inhabitants near this place; yet the *Malayans* come in great Companies when Ships are here; and if they meet with any *Europeans*, they kill them, of what Nation soever they be, not excepting the *Portuguese* themselves.

'Tis

An. 1699. 'Tis but two Years ſince a *Portugueze* Ship riding here, had all the Boats crew cut off as they were Watering; as I was inform'd by the *Dutch*. Here likewiſe is plenty of Fiſh of ſeveral ſorts, which may be catch'd with a Sain; alſo Tortoiſe and Oyſters.

From the North-Eaſt point of this Bay, on the North ſide of the Iſland, the Land trends away North-North-Eaſt for four or five Leagues; afterward North-Eaſt or more Eaſterly; And when you are fourteen or fifteen Leagues to the Eaſtward of *Babao*, you come up with a Point that makes like *Flamborough-Head*, if you are pretty nigh the Land; but if at a diſtance from it on either ſide, it appears like an Iſland. This Point is very remarkable, there being none other like it in all this Iſland. When you are abreaſt of this Point, you will ſee another Point about four Leagues to the Eaſtward; and when you are abreaſt of this latter Point, you will ſee a ſmall Iſland bearing Eaſt or Eaſt by North (according to your diſtance from the Land,) juſt riſing out of the Water: VVhen you ſee it plain, you will be abreaſt of a pretty deep ſandy Bay, which hath a point in the middle, that comes ſloaping from the Mountains, with a curious Valley on each ſide: The ſandy Bay runs from one Valley to the other.

You

You may Sail into this Bay, and anchor a little to the Eastward of the Point in twenty Fathom VVater, half a Mile from the Shore, soft oaze. Then you will be about two Leagues from the VVest-point of the Bay, and about eight Leagues from the small Island before mentioned, which you can see pretty plain bearing East-North-East a little Northwardly. Some other marks are set down in the foregoing Chapter. In this sandy Bay you will find fresh VVater in two or three places. At Spring-tides you will see many riplings, like Sholes; but they are only Eddies caused by the two points of the Bay.

An. 1699.

VVe saw Smoaks all day up in the Mountains, and Fires by Night, at certain places, where we supposed the Natives lived, but saw none of them.

The Tides ran between the two points of the Bay, very strong and uncertain: Yet it did not rise and fall above nine Foot upon a Spring-tide: But it made great riplings and a roaring Noise; whirling about, like Whirlpools. VVe had constantly eddy Tides under the Shore, made by the points on each side of the Bay.

VVhen you go hence to the Eastward, you may pass between the small Island, and *Timor*; and when you are five or six Leagues to the Eastward of the small Island,

An. 1699. you will ſee a large Valley to the Eaſtward of you; then running a little further, you may ſee Houſes on the Bay: You may luff in, but anchor not till you go about the next point. Then you will ſee more Houſes, where you may run into twenty or thirty Fathom, and anchor right againſt the Houſes, neareſt the VVeſt end of them. This place is called *Laphao*. It is a *Portugueze* Settlement, about ſixteen Leagues from the Watering-bay.

There are in it about forty or fifty Houſes, and one Church. The Houſes are mean and low, the Walls generally made of Mud or watled, and their ſides made up with Boards: They are all thatcht with Palm or Palmeto-Leaves. The Church alſo is very ſmall: The Eaſt-end of it is boarded up to the top; but the ſides and the Weſt-end are only boarded three or four foot high; the reſt is all open: There is a ſmall Altar in it, with two Steps to go up to it, and an Image or two; but all very mean. 'Tis alſo thatch'd with Palm or Palmeto-Leaves. Each Houſe has a Yard belonging to it, fenced about with wild Canes nine or ten Foot high. There is a Well in each Yard, and a little Bucket with a String to it to draw Water withal. There is a Trunk of a Tree made hollow, placed in each Well, to keep the Earth from falling

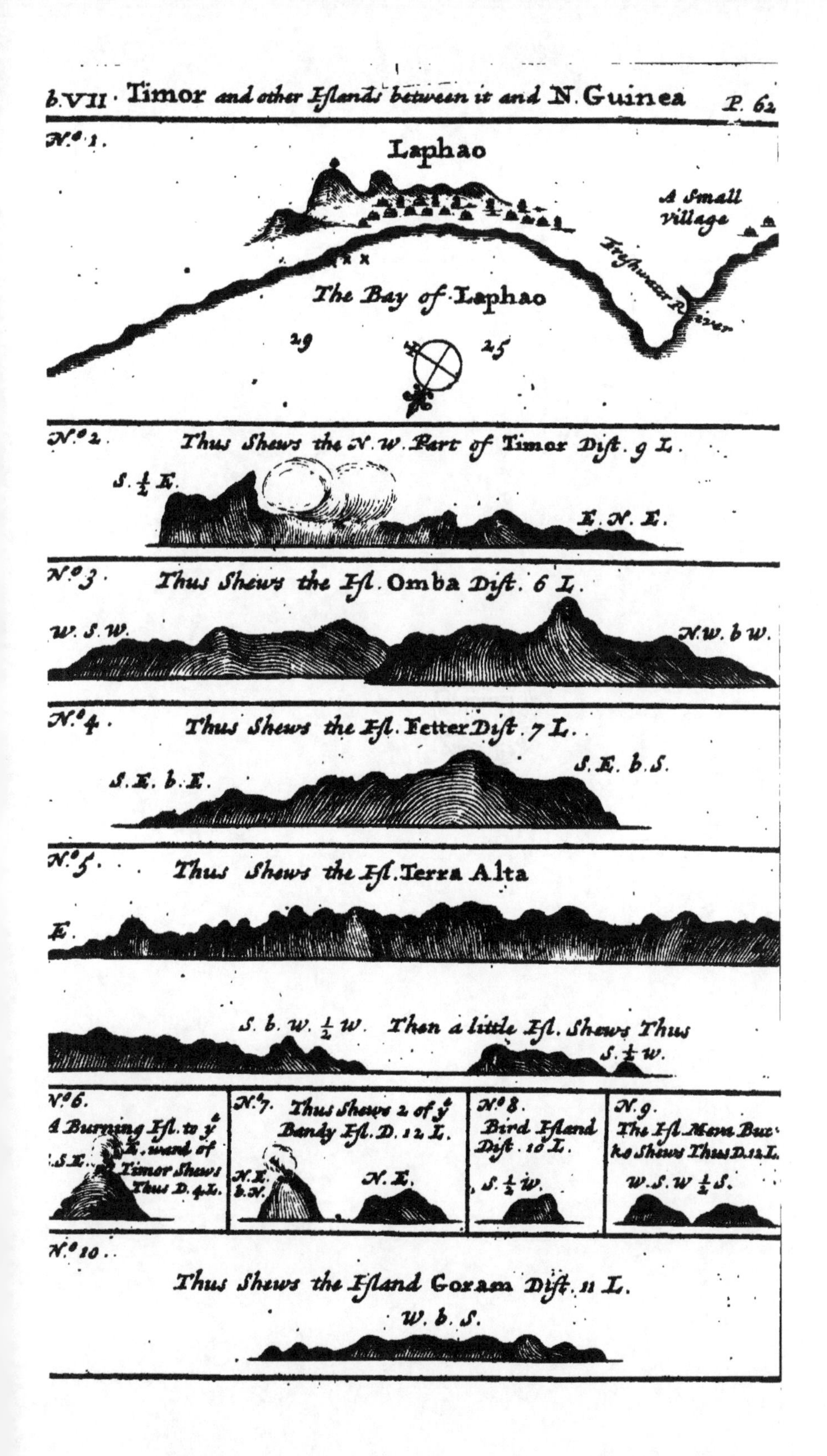

N.° 1.
Laphao
A Small Village
Freshwater River
The Bay of Laphao
29
25
N.° 2
Thus Shews the N.W. Part of Timor Dist. 9 L.
S. ½ E.
E. N. E.
N.° 3.
Thus Shews the Isl. Omba Dist. 6 L.
W. S. W.
N.W. b W.
N.° 4.
Thus Shews the Isl. Fetter Dist. 7 L.
S. E. b. E.
S. E. b. S.
N.° 5.
Thus Shews the Isl. Terra Alta
E.
S. b. W. ½ W.
Then a little Isl. Shews Thus
S. ½ W.
N.° 6.
A Burning Isl. to ye E. ward of Timor Shews Thus D. 4 L.
N.° 7.
Thus Shews 2 of ye Bandy Isl. D. 12 L.
N. E. b. N.
N. E.
N.° 8.
Bird Island Dist. 10 L.
S. ½ W.
N. 9.
The Isl. Mano Burko Shews Thus D. 12 L.
W. S. W ½ S.
N.° 10.
Thus Shews the Island Goram Dist. 11 L.
W. b. S.

ling in. Round the Yards there are many *An.* 1699.
Fruit-trees planted; as Coco-nuts, Tamarins and Toddy-trees.

They have a ſmall Hovel by the Seaſide, where there are ſix ſmall old Iron Guns ſtanding on a decayed Platform, in rotten Carriages. Their Vents are ſo big, that when they are fired, the ſtrength of the Powder flying out there, they give but a ſmall Report, like that of a Musket. This is there Court of Guard; and here were a few armed-men watching all the time we lay here.

The Inhabitants of the Town, are chiefly a ſort of *Indians*, of a Copper-colour, with black lank Hair:. They ſpeak *Portugueze*, and are of the *Romiſh* Religion; but they take the Liberty to eat Fleſh when they pleaſe. They value themſelves on the account of their Religion and deſcent from the *Portugueze*; and would be very angry, if a Man ſhould ſay they are not *Portugueze:* Yet I ſaw but three White Men here, two of which were *Padres.* There are alſo a few *Chineſe* living here. It is a place of pretty good Trade and Strength, the beſt on this Iſland, *Porta-Nova* excepted. They have three or four ſmall Barks belonging to the place; with which they trade chiefly about the Iſland with the Natives, for Wax, Gold, and Sandall-wood. Sometimes

An. 1699. times they go to *Batavia*, and fetch *European* Commodities, Rice, *&c.*

The *Chinese* trade hither from *Macao*; and I was informed that about twenty Sail of ſmall Veſſels come from thence hither every Year. They bring courſe Rice, adulterated Gold, Tea, Iron, and Iron-tools, Porcellane, Silks, *&c.* They take in exchange pure Gold, as 'tis gathered in the Mountains, Bees-wax, Sandall-wood, Slaves, *&c.* Sometimes alſo here comes a Ship from *Goa*. Ships that trade here, begin to come hither the latter end of *March*; and none ſtay here longer than the latter end of *Auguſt*. For ſhould they be here while the North-North-Weſt Monſoon blows, no Cables nor Anchors would hold them; but they would be driven aſhore and daſh'd in pieces preſently. But from *March* till *September*, while the South-South-Eaſt Monſoon blows, Ships ride here very ſecure; For then, though the VVind often blows hard, yet 'tis off Shore; ſo that there is very ſmooth VVater, and no fear of being driven aſhore; And yet even then they moor with three Cables; two towards the Land, Eaſtward and Weſtward; and the third right off to Seaward.

As this is the ſecond place of Traffick, ſo 'tis in Strength the ſecond place the *Portugueze* have here, though not capable of of refifting a hundred Men: For the Pirates that were at the *Dutch* Fort, came hither alfo; and after they had fill'd their VVater, and cut Fire-wood, and refresh'd themfelves, they plunder'd the Houfes, fet them on fire, and went away. Yet I was told, that the *Portugueze* can draw together five or fix hundred Men in twenty-four Hours time, all armed with Hand-Guns, Swords and Piftols; but Powder and Bullets are fcarce and dear. The chief Perfon they have on the Ifland, is named *Antonio Henriquez*; They call him ufually by the Title of Captain *More* or *Maior*. They fay he is a white Man, and that he was fent hither by the Vice-Roy of *Goa*. I did not fee him; for he lives, as I was informed, a great way from hence, at a place call'd *Porta Nova*, which is at the Eaft-end of the Ifland, and by report is a good Harbour; but they fay, that this Captain *More* goes frequently to Wars in Company with the *Indians* that are his Neighbours and Friends, againft other *Indians* that are their Enemies. The next Man to him is *Alexis Mendofa*; he is a Lieutenant, and lives fix or feven Miles from hence, and rules this part of the Country. He is a little Man of the *Indian*-Race, Copper-coloured, with black lank Hair. He fpeaks both the *Indian* and *Portuguefe* Languages; is a Roman Catho-

An. 1699. liek, and seems to be a civil brisk Man. There is another Lieutenant at *Laphao*; who is also an *Indian*; speaks both his own and the *Portuguese* Language very well; is old and infirm, but was very courteous to me.

They boast very much of their Strength here, and say they are able at any time to drive the *Dutch* away from the Island, had they Permission from the King of *Portugal* so to do. But though they boast thus of their Strength, yet really they are very weak; for they have but a few small Arms, and but little Powder: They have no Fort, nor Magazine of Arms; nor does the Vice-Roy of *Goa* send them any now: For though they pretend to be under the King of *Portugal*, they are a sort of lawless People, and are under no Government. It was not long since the Vice-Roy of *Goa* sent a Ship hither, and a Land-Officer to remain here: But Captain *More* put him in Irons, and sent him aboard the Ship again; telling the Commander, that he had no occasion for any Officers; and that he could make better Officers here, than any that could be sent him from *Goa*: And I know not whether there has been any other Ship sent from *Goa* since: So that they have no Supplies from thence: Yet they need not want Arms and Ammunition, seeing they Trade

to *Batavia*. However, they have Swords
and Lances as other *Indians* have; and tho' they are Ambitious to be call'd *Portugueze*, and value themselves on their Religion, yet most of the Men and all the Women that live here, are *Indians*; and there are very few right *Portugueze* in any part of the Island. However of those that call themselves *Portugueze*, I was told there are some thousands; and I think their strength consists more in their Numbers than in good Arms or Discipline.

The Land from hence trends away East by North about 14 Leagues, making many points and sandy Bays, where Vessels may Anchor.

Fourteen Leagues East from *Laphao*, there is a small Harbour called *Ciccale* by the *Portuguese*, and commended by them for an excellent Port; but it is very small, has a narrow Entrance, and lies open to Northerly Winds: Though indeed there are two Ledges of Rocks, one shooting out from the West Point, and the other from the East Point, which break off the Sea; for the Rocks are dry at low Water. This Place is about 60 Leagues from the South-west end of the Island.

The whole of this Island *Timor*, is a very uneven rough Country, full of Hills and small Valleys. In the middle of it there runs a Chain of high Mountains, almost from one end to the other. It is indifferently well watered (even in the dry times) with small Brooks and Springs, but no great Rivers; the Island being but narrow, and such a Chain of Mountains in the middle, that no Water can run far; but, as the Springs break out on one side or other of the Hills, they make their nearest Course to the Sea. In the wet Season, the Valleys and low Lands by the Sea are over-flown with Water; and then the small Drills that run into the Sea, are great Rivers; and the Gulleys, which are dry for three or four Months before, now discharge an impetuous Torrent. The low Land by the Sea-side, is for the most part friable, loose, sandy Soil; yet indifferently fertile and cloathed with Woods. The Mountains are checquered with Woods, and some Spots of Savannahs: Some of the Hills are wholly covered with tall, flourishing Trees; others but thinly; and these few Trees that are on them, look very small, rusty and withered; and the spots of Savannahs among them, appear rocky and barren. Many of the Mountains are rich in Gold, Copper, or both: The Rains wash the Gold out of the Mountains, which the Natives pick up in the adjacent Brooks, as the *Spaniards* do in *America*: How they get the Copper, I know not.

The

The Trees that grow naturally here, are of divers sorts; many of them wholly unknown to me; but such as I have seen in *America* or other places, and grow here likewise, are these, *viz.* Mangrove, white, red and black; Maho, Calabash, several sorts of the Palm-kind; The Cotton-trees are not large, but tougher than those in *America*: Here are also Locust-trees of two or three sorts, bearing Fruit, but not like those I have formerly seen: These bear a large white Blossom, and yield much Fruit, but it is not sweet.

Cana-fistula-trees are very common here; the Tree is about the bigness of our ordinary Apple Trees; their Branches not thick, nor full of Leaves. These and the before-mentioned, blossom in *October* and *November*; the Blossoms are much like our Apple-Tree Blossoms, and about that bigness: At first they are red; but before they fall off, when spread abroad, they are white; so that these Trees in their Season appear extraordinarily pleasant, and yield a very fragrant smell. VVhen the Fruit is ripe, it is round and about the bigness of a Man's Thumb; of a dark brown Colour, inclining to red, and about two foot or two foot and half long. We found many of them under the Trees, but they had no Pulp in them. The Partitions in the middle, are much

An. 1699. at the ſame diſtance with thoſe brought to *England*, of the ſame Subſtance, and ſuch ſmall flat Seeds in them: But whether they be the true *Cana-fiſtula* or no, I cannot tell, becauſe I found no black Pulp in them.

The *Calabaſhes* here are very prickly: The Trees grow tall and tapering; whereas in the *Weſt-Indies* they are low and ſpread much abroad.

Here are alſo Wild *Tamarind*-trees, not ſo large as the true; though much reſembling them both in the Bark and Leaf.

Wild Fig-trees here are many, but not ſo large as thoſe in *America*. The Fruit grows, not on the Branches ſingly, like thoſe in *America*, but in Strings and Cluſters, forty or fifty in a cluſter, about the Body and great Branches of the Tree, from the very Root up to the Top. Theſe Figs are about the bigneſs of a Crab-Apple, of a Greeniſh Colour, and full of ſmall white Seeds; they ſmell pretty well, but have no Juice or Taſte; they are ripe in *November*.

Here likewiſe grows *Sandal*-wood, and many more ſorts of Trees fit for any uſes. The talleſt among them, reſemble our Pines; they are Streight and Clear-bodied, but not very thick; the inſide is reddiſh near the Heart, and hard and Ponderous.

The

Of the Palm-kind there are three or four ſorts; two of which kinds I have not ſeen any where but here. Both ſorts are very large, and tall. The firſt ſort had Trunks of about ſeven or eight Foot in Circumference, and about eighty or ninety Foot high. Theſe had Branches at the top like Coco-nut-Trees, and their Fruit like Coco-nuts, but ſmaller: The Nut was of an Oval form, and about the bigneſs of a Ducks Egg: The ſhell black and very hard. 'Twas almoſt full of Kernel, having only a ſmall empty ſpace in the middle, but no Water as Coco-nuts have. The Kernel is too hard to be eaten. The Fruit ſomewhat reſembles that in *Brazil* formerly mentioned. The husk or outſide of the Fruit, was very Yellow, ſoft and pulpy, when ripe; and full of ſmall Fibres; and when it fell down from the Tree, would maſh and ſmell unſavory.

The other ſort was as big and tall as the former; the Body growing ſtreight up without Limbs, as all Trees of the Palm-kind do: But inſtead of a great many long green Branches growing from the head of the Tree, theſe had ſhort Branches about the bigneſs of a Mans Arm, and about a Foot long; each of which ſpread it ſelf into a great many ſmall tough twigs, that hung full of Fruit like

An. 1699. ſo many Ropes of Onions. The Fruit was as big as a large Plumb; and every Tree had ſeveral Buſhels of Fruit. The Branches that bore this Fruit, ſprouted out at about fifty or ſixty Foot heighth from the ground. The trunk of the Tree was all of one bigneſs, from the Ground to that heighth; but from thence it went tapering ſmaller and ſmaller to the top, where it was no bigger than a Mans Leg, ending in a Stump: And there was no Green about the Tree, but the Fruit; ſo that it appeared like a dead Trunk.

Beſides Fruit-Trees, here were many ſorts of tall Streight-bodied Timber-Trees; one ſort of which, was like Pine. Theſe grow plentifully all round the Iſland by the Sea-ſide, but not far within Land. 'Tis hard Wood, of a reddiſh Colour, and very ponderous.

The Fruits of this Iſland, are *Guavoes*, *Mangoes*, *Jaca's*, *Coco-nuts*, *Plantains*, *Bonanoes*, *Pine-Apples*, *Citrons*, *Pomegranates*, *Oranges*, *Lemons*, *Limes*, *Musk-Melons*, *Water-Melons*, *Pumkins*, &c. Many of theſe have been brought hither by the *Dutch* and *Portugueze*; and moſt of them are ripe in *September* and *October*. There were many other excellent Fruits, but not now in Seaſon; as I was inform'd both by *Dutch* and *Portugueze*.

Here

Here I met with an Herb, which in the *West-Indies* we call *Calalaloo.* It grows wild here. I eat of it ſeveral times, and found it as pleaſant and wholeſome as Spinage. Here are alſo Purſly, Sampier, *&c. Indian* Corn thrives very well here, and is the common Food of the Iſlanders; though the *Portugueze* and their Friends ſow ſome Rice, but not half enough for their ſubſiſtence.

The Land-Animals are Buffaloes, Beeves, Horſes, Hogs, Goats, Sheep, Monkeys, Guanoes, Lizards, Snakes, Scorpions, Centumpees, *&c.* Beſide the tame Hogs and Buffaloes, there are many wild all over the Country, which any may freely kill. As for the Beeves, Horſes, Goats and Sheep, it is probable they were brought in by the *Portugueze* or *Dutch*; eſpecially the Beeves; for I ſaw none but at the *Dutch* Fort *Concordia.*

We alſo ſaw Monkeys, and ſome Snakes. One ſort yellow, and as big as a Mans Arm, and about four Foot long: Another ſort no bigger than the Stem of a Tobacco-pipe, about five Foot long, green all over his Body, and with a flat red head as big as a Mans Thumb.

The Fowls are Wild Cocks and Hens, Eagles, Hawks, Crows, two ſorts of Pidgeons, Turtle-doves, three or four ſorts of Parrots, Parrakites, Cockatoes,

Black-

An. 1699. Black-birds; besides a multitude of smaller Birds of diverse Colours, whose charming Musick makes the Woods very pleasant. One sort of these pretty little Birds my Men call'd the Ringing-bird; because it had six Notes, and always repeated all his Notes twice one after another; beginning high and shrill, and ending low. This Bird was about the bigness of a Lark, having a small sharp black Bill, and blew Wings; the Head and Breast were of a pale red, and there was a blew streak about its Neck. Here are also Sea or Water-Fowls, as Men of War-Birds, Boobies, Fishing-hawks, Herons, Goldens, Crab-catchers, *&c.* The tame Fowl are Cocks, Hens, Ducks, Geese; the two last sorts I only saw at the *Dutch* Fort; of the other sort there are not many but among the *Portugueze.* The Woods abound with Bees, which make much Honey and Wax.

The Sea is very well stock'd with Fish of diverse sorts, *viz.* Mullets, Bass, Breames, Snooks, Mackarel, Parracoots, Gar-fish, Ten-pounders, Scuttle-fish, String-rays, Whip-rays, Rasperages, Cockle-merchants, or Oyster-crackers, Cavallies, Conger-Eels, Rock-fish, Dog-fish, *&c.* The Rays are so plentiful, that I never drew the Sain but I catch'd some of them; which we Salted and Dryed. I caught one whose tail

Tail was thirteen Foot long. The *Cockle-Merchants* are ſhaped like Cavallies, and about their bigneſs. They feed on Shell-fiſh, having two very hard, thick, flat Bones in their Throat, with which they break in pieces the Shells of the Fiſh they ſwallow. We always find a great many Shells in their Maws, cruſhed in pieces. The Shell-fiſh, are Oyſters of three ſorts, *viz.* Long-Oyſters, Common-Oyſters, growing upon Rocks in great abundance, and very Flat; and another ſort of large Oyſters, Fat and Crooked; the Shell of this, not eaſily to be diſtinguiſhed from a Stone. Three or four of theſe Roaſted, will ſuffice a Man for one Meal. Cockles, as big as a Mans Head; of which two or three are enough for a Meal; they are very Fat and Sweet. Craw-fiſh, Shrimps, *&c.* Here are alſo many green Turtle, ſome Alligators and Grand-piſces, *&c.*

The Original Natives of this Iſland, are *Indians*, they are of a middle Stature, Streight-bodied, Slender-limb'd, Long-viſag'd; their Hair black and lank; their Skins very ſwarthy. They are very dextrous and nimble, but withal lazy in the higheſt degree. They are ſaid to be dull in every thing but Treachery and Barbarity. Their Houſes are but low and mean, their cloathing only a ſmall Cloath about their middle; but ſome of them

An. 1699. for Ornament have frontlets of Mother of Pearl, or thin pieces of Silver or Gold, made of an Oval form, of the breadth of a Crown-piece, curiously notched round the edges; Five of these placed one by another a little above the Eye-brows, making a sufficient Guard and Ornament for their Fore-head. They are so thin, and placed on their Fore-heads so artificially, that they seem riveted thereon: And indeed the Pearl-Oyster-shells make a more splendid Show, than either Silver or Gold. Others of them have Palmeto-caps made in diverse forms.

As to their Marriages, they take as many Wives as they can maintain; and sometimes they sell their Children to purchase more Wives. I enquir'd about their Religion, and was told they had none. Their common subsistence is by *Indian* Corn, which every Man plants for himself. They take but little pains to clear their Land; For in the Dry time they set Fire to the withered Grass and Shrubs, and that burns them out a Plantation for the next wet Season. What other Grain they have, beside *Indian* Corn, I know not. Their Plantations are very mean; for they delight most in Hunting; and here are wild Buffaloes and Hogs enough, though very shy, because of their so frequent Hunting.

They

They have a few Boats and ſome Fiſhermen. Their Arms are Lances, thick round ſhort Truncheons and Targets; with theſe they Hunt and kill their Game, and their Enemies too; for this Iſland is now divided into many Kingdoms, and all of different Languages; though in their Cuſtoms and manner of living, as well as Shape and Colour, they ſeem to be of one Stock.

The chiefeſt Kingdoms are *Cupang*, *Amabie*, *Lortribie*, *Pobumbie*, *Namquimal*; the Iſland alſo of *Anamabao* or *Anabao*, is a Kingdom. Each of theſe hath a Sultan, who is Supreme in his Province and Kingdom, and hath under him ſeveral *Raja's* and other inferiour Officers. The Sultans for the moſt part are Enemies to each other; which Enmities are fomented and kept up by the *Dutch*, whoſe Fort and Factory is in the Kingdom of *Cupang*; and therefore the Bay near which they are ſettled, is commonly called *Cupang*-Bay. They have only as much Ground as they can keep within reach of their Guns; yet this whole Kingdom is at peace with them; and they freely trade together; as alſo with the Iſlanders on *Anabao*, who are in Amity as well with the Natives of *Cupang*, as with the *Dutch* reſiding there; but they are implacable Enemies to thoſe of *Amabie*, who are their next

An. 1699. next Neighbours, and in Amity with the *Portugueze*; as are alſo the Kingdoms of *Pobumbie*, *Namquimal* and *Lortribie*. It is very probable, that theſe two *European* Settlements on this Iſland, are the greateſt occaſion of their continued Wars. The *Portugueſe* vaunt highly of their Strength here, and that they are able at pleaſure to rout the *Dutch*, if they had Authority ſo to do from the King of *Portugal*; and they have written to the Vice-Roy of *Goa* about it: And though their Requeſt is not yet granted, yet (as they ſay) they live in expectation of it. Theſe have no Forts, but depend on their Alliance with the Natives: And indeed they are already ſo mixt, that it is hard to diſtinguiſh whether they are *Portugueſe* or *Indians*. Their Language is *Portugueſe*; and the Religion they have, is *Romiſh*. They ſeem in Words to acknowledge the King of *Portugal* for their Sovereign; yet they will not accept of any Officers ſent by him. They ſpeak indifferently the *Malayan* and their own native Languages, as well as *Portugueſe*; and the chiefeſt Officers that I ſaw, were of this ſort; neither did I ſee above three or four white Men among them; and of theſe, two were Prieſts. Of this mixt Breed there are ſome thouſands; of whom ſome have ſmall Arms of their own, and know how

to

to uſe them. The chiefeſt Perſon (as I before ſaid) is called Captain *More* or *Mayor*: He is a white Man, ſent hither by the Vice-Roy of *Goa*, and ſeems to have great Command here. I did not ſee him; for he ſeldom comes down. His Reſidence is at a place called *Porta Nova*; which the people at *Laphao* told me was a great way off; but I could not get any more particular account. Some told me that he is moſt commonly in the Mountains, with an Army of *Indians*, to guard the Paſſes between them and the *Cupangayans*, eſpecially in the dry Times. The next Man to him is *Alexis Mendoſa*: He is a right *Indian*, ſpeaks very good *Portugueſe*, and is of the *Romiſh* Religion. He lives five or ſix Miles from the Sea, and is called the Lieutenant. (This is he whom I call Governour, when at *Laphao*.) He commands next to Captain *More*, and hath under him another at this Fort (at the Sea-ſide) if it may be ſo called. He alſo is called Lieutenant, and is an *Indian Portugueſe*.

Beſides this Mungrel-Breed of *Indians* and *Portugueſe*, here are alſo ſome *China*-Men, Merchants from *Maccao*: They bring hither courſe Rice, Gold, Tea, Iron-work, Porcelane, and Silk both wrought and raw: They get in exchange pure Gold as it is here gather'd, Bees-wax, Sandal-

An. 1699. Sandal-Wood, Coire, *&c.* It is said there are about twenty small *China* Vessels come hither every Year from *Maccao*; and commonly one Vessel a Year from *Goa*, which brings *European* Commodities and Callicoes, Muslins, *&c.* Here are likewise some small Barks belonging to this Place, that Trade to *Batavia*, and bring from thence both *European* and *Indian* Goods and Rice. The Vessels generally come here in *March*, and stay till *September*.

The *Dutch*, as I before said, are setled in the Kingdom of *Cupang*, where they have a small neat Stone Fort. It seems to be pretty strong; yet, as I was informed, had been taken by a *French* Pirate about two Years ago: The *Dutch* were used very barbarously, and ever since are very jealous of any Strangers that come this way; which I my self experienced. These depend more on their own Strength than on the Natives their Friends; having good Guns, Powder, and Shot enough on all occasions, and Soldiers sufficient to manage the Business here, all well disciplin'd and in good order; which is a thing the *Portuguese* their Neighbours are altogether destitute of, they having no *European* Soldiers, few Arms, less Ammunition, and their Fort consisting of no more than six bad Guns planted against the

the Sea, whose Touch-holes (as was before observed) are so enlarg'd by time, that a great part of the strength of the Powder flies away there; And having Soldiers in pay, the Natives on all occasions are hired; and their Government now is so loose, that they will admit of no more Officers from *Portugal* or *Goa*. They have also little or no supply of Arms or Ammunition from thence, but buy it as often as they can, of the *Dutch*, *Chinese*, &c. So that upon the whole it seems improbable that they should ever attempt to drive out the *Dutch*, for fear of loosing themselves, notwithstanding their boasted Prowess and Alliance with the Natives: And indeed, as far as I could learn, they have business enough to keep their own present Territories from the incursions of the *Cupangayans*; who are Friends to the *Dutch*, and whom doubtless the *Dutch* have ways enough to preserve in their Friendship: besides that they have an inveterate Malice to their Neighbours, insomuch that they kill all they meet, and bring away their Heads in Triumph. The great Men of *Cupang* stick the Heads of those they have killed, on Poles; and set them on the tops of their Houses; and these they esteem above all their other Riches. The inferiour sort bring the Heads of

An. 1699 those they kill, into Houses made for that purpose; of which there was one at the *Indian* Village near the Fort *Concordia*, almost full of Heads, as I was told. I know not what encouragement they have for their inhumanity.

The *Dutch* have always two Sloops belonging to their Fort; in these they go about the Island, and Trade with the Natives; and, as far as I could learn, they Trade indifferently with them all. For though the Inland people are at war with each other, yet those by the Sea-side seem to be little concerned; and, generally speaking the *Malayan* Language, are very sociable and easily induced to Trade with those that speak that Language; which the *Dutch* here always learn; Besides, being well acquainted with the Treachery of these People, they go well arm'd among them, and are very vigilant never to give them an opportunity to hurt them; and it is very probable that they supply them with such Goods, as the *Portugueze* cannot.

The *Malayan* Language, as I have before said, is generally spoken amongst all the Islands hereabouts. The greater the Trade is, the more this Language is spoken: In some it is become their only Language; in others it is but little spoken, and that by the Sea-side only. VVith this

this Language the *Mahometan* Religion did ſpread it ſelf, and was got hither before any *European* Chriſtians came: But now, though the Language is ſtill uſed, the *Mahometan* Religion falls, where-ever the *Portugueze* or *Dutch* are ſettled; unleſs they be very weak, as at *Solor* and *Ende*, where the chief Language is *Malayan*, and the Religion Mahometaniſm; though the *Dutch* are ſettled at *Solor*, and the *Portugueze* at the Eaſt end of the Iſland *Ende*, at a place called *Lorantuca*; which, as I was informed, is a large Town, hath a pretty ſtrong Fort and ſafe Harbour. The chief Man there (as at *Timor*) is called Captain *More*, and is as abſolute as the other. Theſe two principal Men are Enemies to each other; and by their Letters and Meſſages to *Goa*, inveigh bitterly againſt each other; and are ready to do all the ill Offices they can; yet neither of them much regards the Vice-Roy of *Goa*, as I was inform'd.

L' Orantuca is ſaid to be more populous, than any Town on *Timor*; the Iſland *Ende* affording greater plenty of all manner of Fruit, and being much better ſupplied with all Neceſſaries, than *Laphao*; eſpecially with Sheep, Goats, Hogs, Poultrey, &c. but it is very dangerous getting into this Harbour, becauſe of the violent Tides, between the Iſlands *Ende* and *Solor*:

An. 1699. *lor*. In the middle Channel between *Timor* and the Range of Islands to the Northward of it, whereof *Ende* and *Solor* are two, there runs a constant Current all the Year to the Westward; though near either Shore there are Tides indeed; but the Tide of Flood, which sets West, running eight or nine hours, and the Ebb not exceeding three or four hours, the Tide in some places riseth nine or ten Foot on a Spring.

The Seasons of the Year here at *Timor*, are much the same as in other places in South Latitude. The fair Weather begins in *April* or *May*, and continues to *October*, then the Tornadoes begin to come, but no violent bad Weather till the middle of *December*. Then there are violent West or North-West Winds, with Rain, till towards the middle of *February*. In *May* the Southerly Winds set in, and blow very strong on the North-side of the Island, but fair. There is great difference of Winds on the two sides of the Island: For the Southerly Winds are but very faint on the South-side, and very hard on the North-side; and the bad Weather on the South-side comes in very violent in *October*, which on the North-side comes not till *December*. You have very good Sea and Land-breezes, when the Weather is fair; and may run indifferently to the East

Eaſt or Weſt, as your buſineſs lies. We found from *September* to *December* the Winds veering all round the Compaſs gradually in twenty four hours time; but ſuch a conſtant Weſtern Current, that it's much harder getting to the Eaſt than Weſt at or near Spring Tides: Which I have more than once made tryal off. For weighing from *Babao* at ſix a Clock in the Morning on the 12th inſtant, we kept plying under the Shore till the 20th, meeting with ſuch a Weſtern Current, that we gain'd very little. We had Land and Sea-breezes; but ſo faint, that we could hardly ſtem the Current; and when it was calm between the Breezes, we drove a-Stern faſter than ever we ſailed a-Head.

CHAP. III.

Departure from Timor. *The Islands* Omba *and* Fetter. *A burning Island. Their missing the* Turtle-Isles. Bande-*Isles.* Bird-*Island. They descry the Coast of* New-Guinea. *They Anchor on the Coast of* New-Guinea. *A description of the place, and of a strange Fowl found there. Great quantities of Mackerel. A white Island. They Anchor at an Island called by the Inhabitants* Pulo Sabuda. *A description of it, and its Inhabitants, and Product. The Indians manner of Fishing there. Arrival at* Mabo, *the North-West Cape of* New-Guinea. *A Description of it.* Cockle-*Island. Cockles of seventy-eight pound Weight.* Pidgeon-*Island. The Winds hereabouts. An empty Cockle-shell weighing two hundred fifty-eight Pound.* King William's

William's *Island.* *A Description* *of it.* *Plying on the Coast of* New-Guinea. *Fault of the Draughts.* Providence *Island.* *They cross the Line.* *A Snake pursued by Fish.* *Squally Island.* *The Main of* New-Guinea.

ON the 12th of *December* 1699, we sailed from *Babao*, coasting along the Island *Timor* to the Eastward, towards *New Guinea*. It was the 20th before we got as far as *Laphao*, which is but forty Leagues. We saw black Clouds in the North-West, and expected the Wind from that Quarter above a Month sooner.

That Afternoon we saw the opening between the Islands *Omba* and *Fetter*, but feared to pass through in the Night. At two a Clock in the Morning, it fell calm; and continued so till Noon, in which time we drove with the Current back again South-West six or seven Leagues.

On the 22d, steering to the Eastward to get through between *Omba* and *Fetter*, we met a very strong Tide against us, so that we, although we had a very fresh Gale, yet made way very slowly; yet before Night, got through. By a good Observation we found that the South-East

An. 1699. point of *Omba* lies in Latitude 8 d. 25 m. In my Draughts it's laid down in 8 deg. 10 min. My true courſe from *Babao*, is Eaſt, 25 deg. North, diſtance one hundred eighty three miles. We founded ſeveral times when near *Omba*, but had no ground. On the North-Eaſt point of *Omba* we ſaw four or five Men, and a little further three pretty Houſes on a low point, but did not go aſhore.

At five this Afternoon, we had a Tornado, which yielded much Rain, Thunder and Lightning; yet we had but little Wind. The 24th in the Morning we catched a large Shark, which gave all the Ships Company a plentiful Meal.

The 27th we ſaw the burning Iſland, it lies in Latitude 6 deg. 36 min. South; it is high, and but ſmall. It runs from the Sea a little ſloaping towards the Top; which is divided in the middle into two Peaks, between which iſſued out much Smoak: I have not ſeen more from any Vulcano. I ſaw no Trees; but the North ſide appeared green, and the reſt look'd very barren.

Having paſt the burning Iſland, I ſhap'd my courſe for two Iſlands called *Turtle Iſles*, which lye North-Eaſt by Eaſt a little Eaſterly, and diſtant about fifty Leagues from the burning Iſle. I fearing the Wind might veer to the Eaſtward of the

North,

North, ſteered twenty Leagues North-East, then North-East by East. On the 28th we ſaw two ſmall low Iſlands, called *Luca-parros*, to the North of us. At noon I accounted my ſelf twenty Leagues ſhort of the *Turtle Iſles*.

The next Morning, being in the Latitude of the *Turtle Iſlands*, we look'd out ſharp for them, but ſaw no appearance of any Iſland, till eleven a Clock; when we ſaw an Iſland at a great diſtance. At firſt we ſuppoſed it might be one of the *Turtle Iſles*: But it was not laid down true, neither in Latitude nor Longitude from the *burning Iſle*, nor from the *Luca-parros*, which laſt I took to be a great help to guide me, they being laid down very well from the *Burning Iſle*, and that likewiſe in true Latitude and diſtance from *Omba*: So that I could not tell what to think of the Iſland now in ſight; we having had fair Weather, ſo that we could not paſs by the *Turtle Iſles* without ſeeing them; and This in ſight was much too far off for them. We found Variation 1 deg. 2 min. Eaſt. In the Afternoon I ſteered North-Eaſt by Eaſt for the Iſlands that we ſaw. At two a Clock I went and look'd over the Fore-yard, and ſaw two Iſlands at much greater diſtance than the *Turtle Iſlands* are laid down in my Draughts; one of them was a very high

peak'd

An. 1699. peak'd mountain, cleft at Top, and much like the *burning Island* that we past by, but bigger and higher; the other was a pretty long high flat Island. Now I was certain that these were not the *Turtle Islands*, and that they could be no other than the *Bande-Isles*; yet we steered in, to make them plainer. At three a Clock we discovered another small flat Island to the North-West of the others, and saw a great deal of Smoak rise from the Top of the high Island; At four we saw other small Islands, by which I was now assured that these were the *Bande-Isles* there. At five I altered my course and steered East, and at eight East-South-East; because I would not be seen by the Inhabitants of those Islands in the Morning. We had little Wind all Night; and in the Morning as soon as 'twas Light, we saw another high peak'd Island: At eight it bore South-South-East half East, distance eight Leagues. And this I knew to be *Bird-Isle*. 'Tis laid down in our Draughts in Latitude 5 deg. 9 min. South, which is too far Southerly by twenty seven miles according to our Observation; And the like error in laying down the *Turtle-Islands*, might be the occasion of our missing them.

At night I shortned Sail, for fear of coming too nigh some Islands, that stretch

away

away bending like a half Moon from *Ceram* towards *Timor*, and which in my course I must of necessity pass through. The next Morning betimes, I saw them; and found them to be at a farther distance from *Bird* Island, than I expected. In the Afternoon it fell quite calm; and when we had a little Wind, it was so unconstant, flying from one point to another, that I could not without difficulty get through the Islands where I designed: Besides, I found a Current setting to the Southward; so that it was betwixt five and six in the Evening, before I past through the Islands; and then just weathered little *Watela*, whereas I thought to have been two or three Leagues more Northerly. We saw the day before, betwixt two and three, a Spout but a small distance from us. It fell down out of a black Cloud, that yielded great store of Rain, Thunder and Lightning: This Cloud hovered to the Southward of us for the space of three hours, and then drew to the Westward a great pace; at which time it was that we saw the Spout, which hung fast to the Cloud till it broke; and then the Cloud whirl'd about to the South-East, then to East North-East; where meeting with an Island, it spent it self and so dispersed; and immediately we had a little of the tail of it, having had

An. 1699. had none before. Afterward we saw a Smoak on the Island *Kosiway*, which continued till Night.

On *New-years-day* we first descried the Land of *New-Guinea*, which appear'd to be high Land: And the next day we saw several high Islands on the Coast of *New-Guinea*, and ran in with the main Land. The Shore here lies along East-South-East and West-North-West. It is high even Land, very well cloathed with tall flourishing Trees, which appear'd very green, and gave us a very pleasant Prospect. We ran to the Westward of four mountainous Islands; And in the night had a small Tornado, which brought with it some Rain and a fair Wind. We had fair Weather for a long time; only when near any Land, we had some Tornadoes; but off at Sea, commonly clear Weather; though if in sight of Land, we usually saw many black Clouds hovering about it.

On the 5th and 6th of *January*, we plied to get in with the Land; designing to anchor, fill Water, and spend a little time in searching the Country, till after the change of the Moon: For I found a strong Current setting against us. We anchor'd in 38 Fathom Water, good oazie Ground. We had an Island of a League long without us, about three Miles distant; and we rode from the Main about a Mile.

The

Table VIII. New Guinea

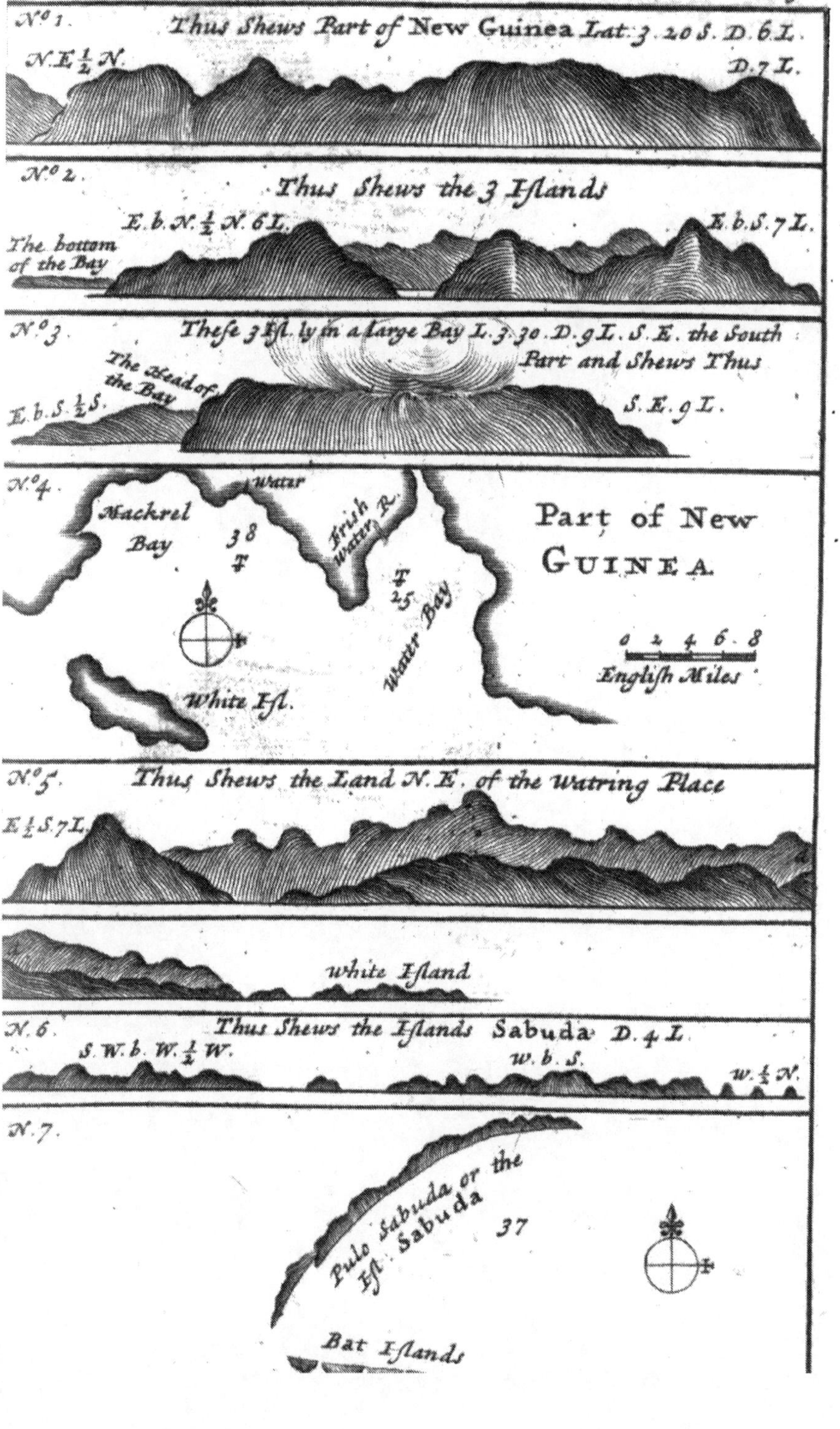

N° 1. Thus Shews Part of New Guinea Lat. 3. 20 S. D. 6 L.

N. E. ½ N.

D. 7 L.

N° 2. Thus Shews the 3 Islands

E. b. N. ½ N. 6 L.

E. b. S. 7 L.

The bottom of the Bay

N° 3. These 3 Isl. ly in a large Bay L. 3. 30. D. 9 L. S. E. the South Part and Shews Thus

The Head of the Bay

E. b. S. ½ S.

S. E. 9 L.

N° 4.

Part of New Guinea

Mackrel Bay

Watter

Frish Watter R.

38

25

Watter Bay

White Isl.

0 2 4 6 8

English Miles

N° 5. Thus Shews the Land N. E. of the Watring Place

E ½ S. 7 L.

White Island

N. 6. Thus Shews the Islands Sabuda D. 4 L.

S. W. b. W. ½ W.

W. b. S.

W. ½ N.

N. 7.

Pulo Sabuda or the Isl. Sabuda

37

Bat Islands

The Eastermost Point of Land seen, bore East by South half South, distance three Leagues: And the Westermost, West-South-West half South, distance two Leagues. So soon as we anchor'd, we sent the Pinnace to look for Water, and try if they could catch any Fish. Afterwards we sent the Yawle another way to see for Water. Before night the Pinnace brought on board several sort of Fruits, that they found in the Woods; such as I never saw before. One of my Men killed a stately Land-Fowl, as big as the largest Dunghil-Cock. It was of a Sky-colour; only in the middle of the Wings was a white Spot, about which were some reddish Spots: On the Crown it had a large Bunch of long Feathers, which appear'd very pretty. His Bill was like a Pidgeons; he had strong Legs and Feet, like Dunghil-Fowls; only the Claws were reddish. His Crop was full of small Berries. It lays an Egg as big as a large Hen's Egg; for our Men climb'd the Tree where it nested, and brought off one Egg. They found Water; and reported that the Trees were large, tall and very thick; and that they saw no sign of People. At night the Yawle came aboard, and brought a wooden Fissgigg, very ingeniously made; the matter of it was a small Cane; They found it by a small Barbecue, where they also saw a shatter'd Canoa. The

An. 1699. The next Morning I sent the Boatswain ashore a fishing, and at one haul he catcht Three hundred fifty-two Mackarels, and about twenty other Fishes; which I caused to be equally divided among all my Company. I sent also the Gunner and chief Mate, to search about if they could find convenient anchoring nearer a Watering-place: By night they brought word that they had found a fine Stream of good Water, where the Boat could come close to, and it was very easie to be fill'd; and that the Ship might anchor as near to it as I pleas'd: So I went thither The next Morning therefore we anchor'd in twenty-five Fathom Water, soft oazie Ground, about a Mile from the River: We got on board three Tun of Water that night; and caught two or three Pike-fish, in shape much like a Parracota, but with a longer Snout, something resembling a Garr, yet not so long. The next day I sent the Boat again for Water, and before night all my Casks were full.

Having fill'd here about fifteen Tuns of Water, seeing we could catch but little Fish, and had no other Refreshments, I intended to sail next day; but finding that we wanted Wood, I sent to cut some; and going ashore to hasten it, at some distance from the place where our Men were, I found a small Cove, where I saw two

Bar-

Fishes taken on the Coast of New Guinea

This Fish fins & tail are blew on y^e edges & red in the middle with blew spots all over y^e Body, but y^e Belly white.

94 .

A Pike Fish Conger on y^e Coast of New Guinea

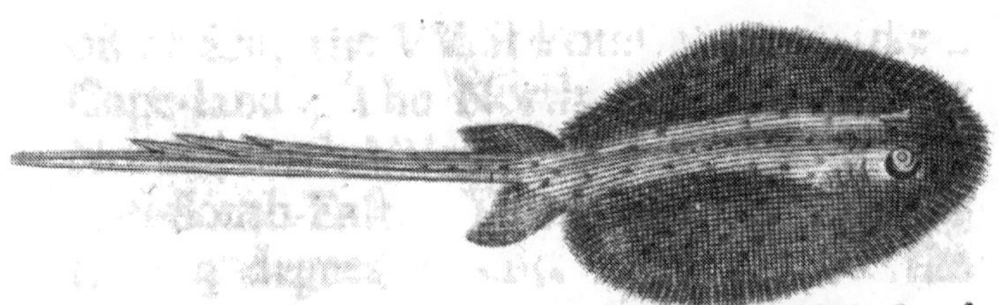

This Fish is a pale red with blew spots on y^e body. the long Tail blew in y^e midle & white on y^e side.

Barbecues, which appear'd not to be above two Months standing: The Sparrs were cut with some sharp Instrument; so that, if done by the Natives, it seems that they have Iron. On the 10th, a little after twelve a-Clock, we weighed and stood over to the North side of the Bay; and at one a-Clock stood out with the Wind at North and North-North-West. At four we past out by a VVhite Island, which I so named from its many white Cliffs, having no name in our Draughts. It is about a League long, pretty high, and very woody: 'Tis about five Miles from the Main, only at the VVest-end it reaches within three Miles of it. At some distance off at Sea, the VVest Point appears like a Cape-land; The North side trends away North-North-VVest, and the East side East-South-East. This Island lies in Latitude 3 degees 4 min. South; and the Meridian Distance from *Babao*, five hundred and twelve Miles East. After we were out to Sea, we plied to get to the Northward; but met with such a strong Current against us, that we got but little. For if the Wind favour'd us in the night, that we got three or four Leagues; we lost it again, and were driven as far astern next Morning; so that we plyed here several Days.

The

An. 1699

The 14th, being paſt a point of Land that we had been three days getting about, we found little or no Current; ſo that having the Wind at North-VVeſt by VVeſt and VVeſt-North-VVeſt, we ſtood to the Northward, and had ſeveral Soundings: At three a-Clock, thirty-eight Fathom; the neareſt part of *New-Guinea* being about three Leagues diſtance: At four, thirty-ſeven; at five, thirty-ſix; at ſix, thirty-ſix; at eight, thirty-three Fathom; Then the Cape was about four Leagues diſtant; ſo that as we ran off, we found our Water ſhallower. We had then ſome Iſlands to the VVeſtward of us, at about four Leagues diſtance.

A little after noon we ſaw Smokes on the Iſlands to the VVeſt of us; and having a fine Gale of VVind, I ſteered away for them: At ſeven a Clock in the Evening we anchored in thirty-five Fathom, about two Leagues from an Iſland; good ſoft oazie Ground. VVe lay ſtill all night, and ſaw Fires aſhore. In the Morning we weighed again, and ran farther in, thinking to have ſhallower VVater; but we ran within a Mile of the Shore, and came to in thirty-eight Fathom, good ſoft holding Ground. While we were under Sail, two Canoas came off within call of us: They ſpoke to us, but we did not underſtand their Language,

nor

nor Signs. VVe wav'd to them to come aboard, and I call'd to them in the *Malayan* Language to do the same; but they would not: Yet they came so nigh us, that we could shew them such things as we had to truck with them; Yet neither would this entice them to come aboard; but they made Signs for us to come ashore, and away they went. Then I went after them in my Pinnace, carrying with me Knives, Beads, Glasses, Hatchets, *&c.* When we came near the Shore, I called to them in the *Malayan* Language: I saw but two Men at first, the rest lying in Ambush behind the Bushes; but assoon as I threw ashore some Knives and other Toys, they came out, flung down their Weapons, and came into the Water by the Boats side, making signs of Friendship by pouring Water on their Heads with one Hand, which they dipt into the Sea. The next day in the Afternoon several other Canoas came aboard, and brought many Roots and Fruits, which we purchas'd.

This Island has no name in our Draughts, but the Natives call it *Pulo Sabuda.* It is about three Leagues long, and two Miles wide, more or less. It is of a good heighth, so as to be seen eleven or twelve Leagues. It is very Rocky; yet above the Rocks there is good yellow

An. 1699. and black Mould; not deep, yet producing plenty of good tall Trees, and bearing any Fruits or Roots which the Inhabitants plant. I do not know all its produce; but what we saw, were Plantains, Coco-Nuts, Pine-Apples, Oranges, Papaes, Potatoes, and other large Roots. Here are also another sort of wild Jaca's, about the bigness of a Mans two Fists, full of Stones or Kernels, which eat pleasant enough when roasted. The Libby Tree grows here in the Swampy Valleys, of which they make Sago Cakes: I did not see them make any, but was told by the Inhabitants that it was made of the Pith of the Tree, in the same manner I have described in my Voyage round the World. They shew'd me the Tree whereof it was made, and I bought about forty of the Cakes. I bought also three or four Nutmegs in their Shell, which did not seem to have been long gathered; but whether they be the growth of this Island or not, the Natives would not tell whence they had them, and seem'd to prize them very much. What Beasts the Island affords, I know not: But here are both Sea and Land-Fowl. Of the first, Boobies and Men of War-Birds are the chief; some Goldens, and small Milk-white Crab-catchers. The Land-fowls are Pidgeons, about the bigness

Vol: III. Part 2
N.
This Fish is of a pale red all parts of it except ye Eye take on ye Coast of New Guinea
Strange & large Batts on I. Pulo Sabuda in New Guinea described Page 99.
This Birds Eye is of a Bright red
Place this Page 99.

neſs of Mountain-Pigeons in *Jamaica*; and Crows about the bigneſs of thoſe in *England*, and much like them; but the inner part of their Feathers are white, and the outſide black; ſo that they appear all black, unleſs you extend the Feathers. Here are large Sky-colour'd Birds, ſuch as we lately kill'd on *New Guinea*; and many other ſmall Birds, unknown to us. Here are likewiſe abundance of Bats, as big as young Coneys; their Necks, Head, Ears and Noſes, like Foxes; their Hair rough; that about their Necks, is of a whitiſh yellow, that on their Heads and Shoulders black; their Wings are four Foot over, from tip to tip: They ſmell like Foxes. The Fiſh are Baſs, Rock-fiſh, and a ſort of Fiſh like Mullets, Old-wives, Whip-rays, and ſome other ſorts that I know not, but no great plenty of any; for 'tis deep Water till within leſs than a Mile of the Shore; then there is a bank of Coral Rocks, within which you have Shoal Water, White clean Sand: So there is no good Fiſhing with the Sain.

This Iſland lies in Latitude 2 deg. 43 min. South, and Meridian diſtance from Port *Babao* on the Iſland *Timor*, four hundred eighty ſix miles. Beſides this Iſland, here are nine or ten other ſmall Iſlands, as they are laid down in the Draughts.

An. 1699

The Inhabitants of this Island are a ſort of very tawny *Indians*, with long black Hair; who in their manners differ but little from the *Mindanayans*, and others of theſe Eaſtern Iſlands. Theſe ſeem to be the chief; For beſides them we ſaw alſo ſhock Curl-pated *New Guinea Negroes*; many of which are Slaves to the others, but I think not all. They are very poor, wear no Cloaths, but have a Clout about their middle, made of the Rinds of the Tops of Palmeto Trees; but the Women had a ſort of Callico Cloaths. Their chief Ornaments are Blue and Yellow-beads, worn about their Wriſts. The Men Arm themſelves with Bows and Arrows, Lances, broad Swords like thoſe of *Mindanao*; their Lances are pointed with Bone. They ſtrike Fiſh very ingeniouſly with Wooden Fiſſ-gigs, and have a very ingenious way of making the Fiſh riſe: For they have a piece of Wood curiouſly carv'd and painted much like a Dolphin (and perhaps other Figures;) theſe they let down into the Water by a Line with a ſmall weight to ſink it; when they think it low enough, they haul the Line into their Boats very faſt, and the Fiſh riſe up after this Figure; and they ſtand ready to ſtrike them when they are near the Surface of the Water. But their chief Livelihood is from their Plantations. Yet they

they have large Boats, and go over to New Guinea, where they get Slaves, fine Parrots, &c. which they carry to Goram and exchange for Callicoes. One Boat came from thence a little before I arriv'd here; of whom I bought some Parrots; and would have bought a Slave, but they would not barter for any thing but Callicoes, which I had not. Their Houses on this side were very small, and seem'd only to be for Necessity; but on the other side of the Island we saw good large Houses. Their Proes are narrow with Outlagers on each side, like other *Malayans.* I cannot tell of what Religion these are; but I think they are not *Mahometans*, by their drinking Brandy out of the same Cup with us without any Scruple. At this Island we continued till the 20th Instant, having laid in store of such Roots and Fruits as the Island afforded.

An. 1699.

On the 20th, at half hour after six in the Morning, I weigh'd, and standing out we saw a large Boat full of Men lying at the North point of the Island. As we passed by, they row'd away towards their Habitations, where we supposed they had withdrawn themselves for fear of us (tho' we gave them no cause of terrour,) or for some differences among themselves.

An. 1699. We stood to the Northward till seven in the Evening; then saw a ripling: and the Water being discoloured, we sounded, and had but twenty two Fathom. I went about and stood to the Westward till two next Morning; then tack'd again, and had these several soundings: At eight in the Evening, twenty two; at ten, twenty five; at eleven, twenty seven; at twelve, twenty eight Fathom; at two in the Morning, twenty six; at four, twenty four; at six, twenty three; at eight, twenty eight; at twelve, twenty two.

We passed by many small Islands, and among many dangerous Shoals, without any remarkable occurrence, till the 4th of *February*, when we got within three Leagues of the North-West Cape of *New Guinea*, called by the *Dutch* Cape *Mabo*. Off this Cape there lies a small woody Island, and many Islands of different Sizes to the North and North-East of it. This part of *New Guinea* is high Land, adorn'd with tall Trees that appeared very Green and Flourishing. The Cape it self is not very high, but ends in a low sharp point; and on either side there appears another such point at equal distances, which makes it resemble a Diamond. This only appears when you are abreast of the middle point; and then you have no ground within three Leagues of the Shore.

In

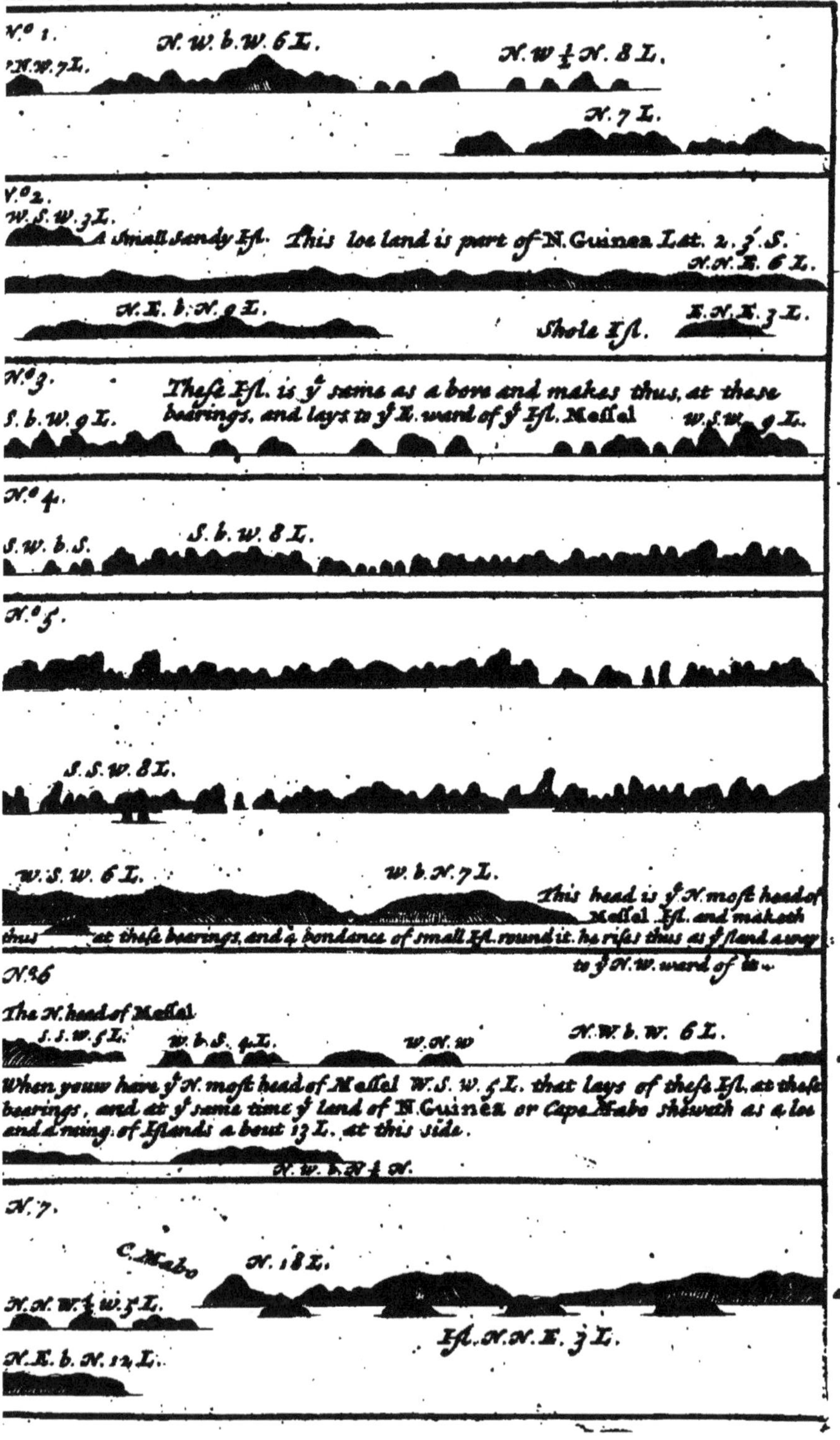

N.° 1.

N.W. 7 L. N.W. b. W. 6 L. N.W ½ N. 8 L.

N. 7 L.

N.° 2.

W. S. W. 3 L. A small sandy Isl. This loe land is part of N. Guinea Lat. 2. 3′. S.

N. N. E. 6 L.

N. E. b. N. 9 L. Shole Isl. E. N. E. 3 L.

N.° 3.

These Isl. is ye same as a bove and makes thus, at these bearings, and lays to ye E. ward of ye Isl. Messel

S. b. W. 9 L. W. S. W. 9 L.

N.° 4.

S. W. b. S. S. b. W. 8 L.

N.° 5.

S. S. W. 8 L.

W. S. W. 6 L. W. b. N. 7 L.

This head is ye N. most head of Messel Isl. and maketh thus at these bearings, and a bondance of small Isl. round it. he rises thus as ye Island away to ye N. W. ward of it.

N.° 6

The N. head of Messel

S. S. W. 5 L. W. b. S. 4 L. W. N. W N. W. b. W. 6 L.

When youw have ye N. most head of Messel W. S. W. 5 L. that lays of these Isl. at these bearings, and at ye same time ye land of N. Guinea or Cape Mabo sheweth as a loe and a ring of Islands a bout 13 L. at this side.

N. W. b. N ½ N.

N. 7.

C. Mabo N. 18 L.

N. N. W. ½ W. 5 L.

Isl. N. N. E. 3 L.

N. E. b. N. 12 L.

In the Afternoon we past by the Cape, and stood over for the Islands. Before it was dark, we were got within a League of the Westermost; but had no ground with fifty Fathom of Line. However fearing to stand nearer in the dark, we tack'd and stood to the East, and plyed all Night. The next Morning we were got five or six Leagues to the Eastward of that Island; and having the Wind Easterly, we stood in to the Northward among the Islands, sounded, and had no ground. Then I sent in my Boat to sound, and they had ground with fifty Fathom near a mile from the Shore. We tack'd before the Boat came aboard again, for fear of a Shoal that was about a mile to the East of that Island the Boat went to; from whence also a Shoal-point stretched out it self till it met the other: They brought with them such a Cockle, as I have mentioned in my Voyage round the World, found near *Celebes*; and they saw many more, some bigger than that which they brought aboard, as they said; and for this reason I named it *Cockle*-Island. I sent them to sound again, ordering them to Fire a Musquet if they found good Anchoring; we were then standing to the Southward, with a fine Breeze. Assoon as they fired, I tack'd and stood in: They told me they had fifty Fathom

An. 1699. when they fired. I tack'd again, and made all the Sail I could to get out, being near ſome Rocky Iſlands and Shoals to Leeward of us. The Breeze increaſed, and I thought we were out of danger; but having a Shole juſt by us, and the VVind falling again, I ordered the Boat to tow us, and by their help we got clear from it. We had a ſtrong Tide ſetting to the Weſtward.

At One a-Clock, being paſt the Shole, and finding the Tide ſetting to the Weſtward, I anchor'd in thirty-five Fathom, courſe Sand, with ſmall Coral and Shells. Being neareſt to *Cockle-Iſland*, I immediately ſent both the Boats thither; one to cut Wood, and the other to fiſh. At four afternoon, having a ſmall Breeze at South-South-Weſt, I made a Sign for my Boats to come aboard. They brought ſome Wood, and a few ſmall Cockles, none of them exceeding ten pound weight; whereas the Shell of the great one weighed ſeventy-eight Pound; but it was now high Water, and therefore they could get no bigger. They alſo brought on board ſome Pidgeons, of which we found plenty on all the Iſlands where we touch'd in theſe Seas. Alſo in many places we ſaw many large Batts, but kill'd none, except thoſe I mention'd at *Pulo Sabuda*. As our Boats came aboard, we weigh'd and made Sail, ſteering East-South-East as long as the Wind held: In the Morning we found we had got four or five Leagues to the East of the place where we weighed. We ſtood to and fro till eleven; and finding that we loſt Ground, anchor'd in forty-two Fathom, courſe gravelly Sand, with ſome Coral. This Morning we thought we ſaw a Sail.

An. 1699

In the Afternoon I went aſhore on a ſmall woody Iſland, about two Leagues from us. Here I found the greateſt number of Pidgeons that ever I ſaw either in the *Eaſt* or *Weſt-Indies*, and ſmall Cockles in the Sea round the Iſland, in ſuch quantities that we might have laden the Boat in an hours time: Theſe were not above ten or twelve pound weight. We cut ſome Wood, and brought off Cockles enough for all the Ship's Company; but having no ſmall Shot, we could kill no Pidgeons. I return'd about four a-Clock; aud then my Gunner and both Mates went thither, and in leſs than three quarters of an Hour they kill'd and brought off ten Pidgeons. Here is a Tide: The Flood ſets Weſt and the Ebb Eaſt; but the latter is very faint, and but of ſmall continuance. And ſo we found it ever ſince we came from *Timor.* The Winds we found Eaſterly, between North-Eaſt and Eaſt-South-Eaſt;

So

An. 1699. So that if these continue, it is impossible to beat farther to the Eastward on this Coast against Wind and Current. These Easterly Winds encreased from the time we were in the Latitude of about 2 deg. South; and as we drew nigher rhe Line, they hung more Easterly. And now being to the North of the Continent of *New Guinea*, where the Coast lies East and West, I find the Trade-wind here at East; which yet in higher Latitudes is usually at North North-West and North-West; and so I did expect them here, it being to the South of the Line.

The 7th in the Morning I sent my Boat ashore on *Pidgeon-Island*, and staid till Noon. In the Afternoon my Men returned, brought twenty-two Pidgeons, and many Cockles; some very large, some small: They also brought one empty Shell, that weigh'd two hundred and fifty-eight Pound.

At four a-Clock we weigh'd, having a small Westerly Wind, and a Tide with us; At seven in the Evening we anchor'd in forty-two Fathom, near *King William's Island*, where I went ashore the next Morning, drank his Majesty's Health, and honour'd it with his Name. It is about two Leagues and a half in length, very high, and extraordinarily well cloathed with Woods. The Trees are of diverse

sorts,

ſorts, moſt unknown to us, but all very green and flouriſhing; many of them had Flowers, ſome white, ſome purple, others yellow; all which ſmelt very fragrantly. The Trees are generally tall and ſtreight-bodied, and may be fit for any uſes. I ſaw one of a clean Body, without Knot or Limb, ſixty or ſeventy Foot high by eſtimation. It was three of my Fathoms about, and kept its bigneſs without any ſenſible decreaſe even to the top. The Mould of the Iſland is black, but not deep; it being very rocky. On the ſides and top of the Iſland, are many Palmeto Trees, whoſe Heads we could diſcern o-ver all the other Trees, but their Bodies we could not ſee.

About one in the Afternoon we weigh-ed and ſtood to the Eaſtward, between the Main and *King William's Iſland*; lea-ving the Iſland on our Larboard ſide, and ſounding till we were paſt the Iſland; and then we had no Ground. Here we found the Flood ſetting Eaſt by North, and the Ebb VVeſt by South. There were Shoals and ſmall Iſlands between us and the Main, which cauſed the Tide to ſet very inconſtantly, and make many whirlings in the VVater; yet we did not find the Tide to ſet ſtrong any way, nor the VVater to riſe much.

On

An. 1699. On the 9th, being to the Eastward of *King William's Island*, we plied all day between the Main and other Islands, having Easterly VVinds and fair weather till seven the next Morning. Then we had very hard Rain till eight, and saw many Sholes of Fish. We lay becalm'd off a pretty deep Bay on *New Guinea*, about twelve or fourteen Leagues wide, and seven or eight Leagues deep, having low Land near its bottom, but high Land without. The Eastermost part of *New Guinea* seen, bore East by South, distant twelve Leagues: Cape *Mabo* West-South-West half South, distant seven Leagues.

At one in the Afternoon it began to rain, and continu'd till six in the Evening; so that having but little Wind and most Calms, we lay still off the formention'd Bay, having *King William's Island* still in sight, though distant by Judgment fifteen or sixteen Leagues West. We saw many Sholes of small Fish, some Sharks, and seven or eight Dolphins; but catcht none. In the Afternoon, being about four Leagues from the Shore, we saw an Opening in the Land, which seem'd to afford good Harbour: In the Evening we saw a large Fire there; and I intended to go in (if Winds and Weather would permit) to get some Acquaintance with the Natives.

Since

N.° 1.

N.N.W. 12 L. N.½ E. 6 L. C. Mabo

N.E. b. E. 7 L. E. b. N. 9 L.

Thus shews Cape Mabo and ye Islands to ye Westward at these Bearing N.N.W. 12 L. also ye low Isl. to ye Eastward of ye Cape at ye Bearing E. b. S.½ S. 7 L.

These are low Islands E. b. S. ½ S. 7 L.

N.° 2.

S.W. b. S. W. b. N. W.N.W. 2 L. N. b. E. 7 L.

N.N.E.½ E.

When you have Cape Mabo S.E. b. E. 5 L. that shews ye Islands to ye Northward of the North Part of N. Guinea at these Bearings & distances.

N.E. b.E.½ E. E.N. E. 10 L. King Willm Island

N.° 3.

S.S.E. 8 L. The Cape of Good Hope S.½ E. 6 L.

Thus shews the Cape of Good Hope at these bearings and dist. and ye land to the E. and Westward

S. b. W. ½ W. 9 L.

N.° 4.

S.E.½ E. Van Scoutens Isl.

Thus shews ye Isl. Providence and van Scoutens. at these Bearings and Dist.

S.½ E. 10 L. The Isl. Providence S.½ W. 3 L.

N.° 5.

S.S.E. 10 L.

Thus shews St Mathias Isl. Dist. from the middle 5 L.

S. S.W ½ W. 7 L.

Since the 4th instant that we passed Cape *Mabo*, to the 12th, we had small Easterly Winds and Calms, so that we anchor'd several times; where I made my Men cut Wood, that we might have a good Stock when a Westerly Wind should present; and so we ply'd to the Eastward, as Winds and Currents would permit; having not got in all above thirty Leagues to the Eastward of Cape *Mabo*. But on the 12th, at four in the Afternoon, a small Gale sprung up at North-East by North, with Rain: At five it shuffled about to North-West, from thence to the South-West, and continued between those two Points a pretty brisk Gale; so that we made Sail and steered away North-East, till the 13th in the Morning, to get about the *Cape of Good Hope*. When 'twas Day, we steer'd North-East half East, then North-East by East till seven a-Clock; and being then seven or eight Leagues off Shore, we steer'd away East; the Shore trending East by South. We had very much Rain all night, so that we could not carry much Sail; yet we had a very steddy Gale. At eight this Morning the VVeather clear'd up, and the VVind decreas'd to a fine Top-gallant Gale, and settled at VVest by South. VVe had more Rain these three Days past, than all the Voyage in so short time. We were now about

An. 1699. about six Leagues from the Land of *New-Guinea*, which appear'd very high; And we saw two Head-lands, about twenty Leagues asunder; the one to the East, and the other to the West, which last is called the *Cape of Good Hope*. We found Variation East 4 deg.

The 15th in the Morning between twelve and two a-Clock, it blew a very brisk Gale at North-West, and look'd very black in the South-West. At two it flew about at once to the South-South-West, and rained very hard. The VVind settled sometime at West-South-West, and we steered East-North-East till three in the Morning: Then the Wind and Rain abating, we steered East half North for fear of coming near the Land. Presently after, it being a little clear, the Man at the Bowsprit-end, call'd out *Land on our Starboard Bow*. VVe lookt out and saw it plain. I presently sounded, and had but ten Fathom soft Ground. The Master, being somewhat scar'd, came running in haste with this News, and said it was best to anchor: I told him no, but sound again: Then we had twelve Fathom; the next Cast, thirteen and a half; the fourth, seventeen Fathom; and then no Ground with fifty Fathom Line. However we kept off the Island, and did not go so fast but that we could see any

other

other danger before we came nigh it. For *An.* 1699. here might have been more Islands not laid down in my Draughts besides This. For I search'd all the Draughts I had, if perchance I might find any Island in the one, which was not in the others; But I could find none near us. When it was day, we were about five Leagues off the Land we saw; but, I believe, not above five Mile or at most two Leagues off it, when we first saw it in the Night.

This is a small Island, but pretty high; I named it *Providence*. About five Leagues to the Southward of this, there is another Island, which is called *William Scouten's Island*, and laid down in our Draughts: It is a high Island, and about twenty Leagues long.

It was by mere Providence that we miss'd the small Island. For had not the Wind come to West-South-West, and blown hard, so that we steered East-North-East; we had been upon it by our course that we steered before, if we could not have seen it. This morning we saw many great Trees and Logs swim by us; which it's probable came out of some great Rivers on the Main.

On the 16th we crossed the Line, and found Variation 6 deg. 26 min. East. The 18th by my observation at noon, we found that we had had a Current setting

to

An. 1699. to the Southward, and probably that drew us in ſo nigh *Scouten's* Iſland. For this twenty-four Hours we ſteered Eaſt by North with a large VVind, yet made but an Eaſt by South half South courſe; though the Variation was not above 7 deg. Eaſt.

The 21ſt we had a Current ſetting to the Northward, which is againſt the true Trade Monſoon, it being now near the full Moon. I did expect it here, as in all other places. VVe had Variation 8 deg. 45 min. Eaſt. The 22d we found but little Current; if any, it ſet to the Southward.

On the 23d in the Afternoon we ſaw two Snakes; and the next Morning another, paſſing by us, which was furiouſly aſſaulted by two Fiſhes, that had kept us Company five or ſix days. They were ſhaped like Mackarel, and were about that bigneſs and length, and of a yellow greeniſh Colour. The Snake ſwam away from them very faſt, keeping his Head above Water; the Fiſh ſnap'd at his Tail; but when he turn'd himſelf, that Fiſh would withdraw, and another would ſnap; ſo that by turns they kept him employed; yet he ſtill defended himſelf, and ſwam away a great pace, till they were out of ſight.

The

An. 1699

The 25th betimes in the Morning, we ſaw an Iſland to the Southward of us, at about fifteen Leagues diſtance. We ſteer'd away for it, ſuppoſing it to be that which the *Dutch* call *Wiſhart's* Iſland; but finding it otherwiſe, I called it *Matthias*; it being that Saints day. This Iſland is about nine or ten Leagues long, Mountainous and Woody, with many Savanna's, and ſome ſpots of Land which ſeem'd to be clear'd.

At 8 in the Evening we lay by; intending, if I could, to anchor under *Matthias* Iſle. But the next Morning ſeeing another Iſland about ſeven or eight Leagues to the Eaſtward of it, we ſteer'd away for it; at noon we came up fair with its South-Weſt-end, intending to run along by it, and Anchor on the South-Eaſt ſide: But the Tornadoes came in ſo thick and hard, that I could not venture in. This Iſland is pretty low and plain, and cloath'd with Wood; the Trees were very green, and appear'd to be large and tall, as thick as they could ſtand one by another. It is about two or three Leagues long, and at the South-VVeſt point there is another ſmall low woody Iſland, about a mile round; and about a mile from the other. Between them there runs a riff of Rocks, which joyns them. (The biggeſt, I named *Squally Iſland*.)

An. 1699. Seeing we could not anchor here, I stood away to the Southward, to make the Main. But having many hard Squalls and Tornadoes, we were often forced to hand all our Sails and steer more Easterly to go before it. On the 26th at four a Clock it clear'd up to a hard Sky, and a brisk settled Gale; then we made as much Sail as we could. At five it clear'd up over the Land, and we saw, as we thought, Cape *Solomaswer* bearing South-South-East distance ten Leagues. VVe had many great Logs and Trees swimming by us all this Afternoon, and much Grass; we steered in South-South-East till six, then the VVind slackned, and we stood off till seven, having little VVind: then we lay by till ten, at which time we made Sail, and steer'd away East all Night. The next Morning, as soon as it was light, we made all the Sail we could, and steer'd away East-South-East, as the Land lay; being fair in sight of it, and not above seven Leagues distance. We past by many small low woody Islands which lay between us and the Main, not laid down in our Draughts. VVe found Variation 9 deg. 50 min. East.

The 28th we had many violent Tornadoes, VVind, Rain, and some Spouts; and in the Tornadoes the VVind shifted. In the Night we had fair VVeather, but

more

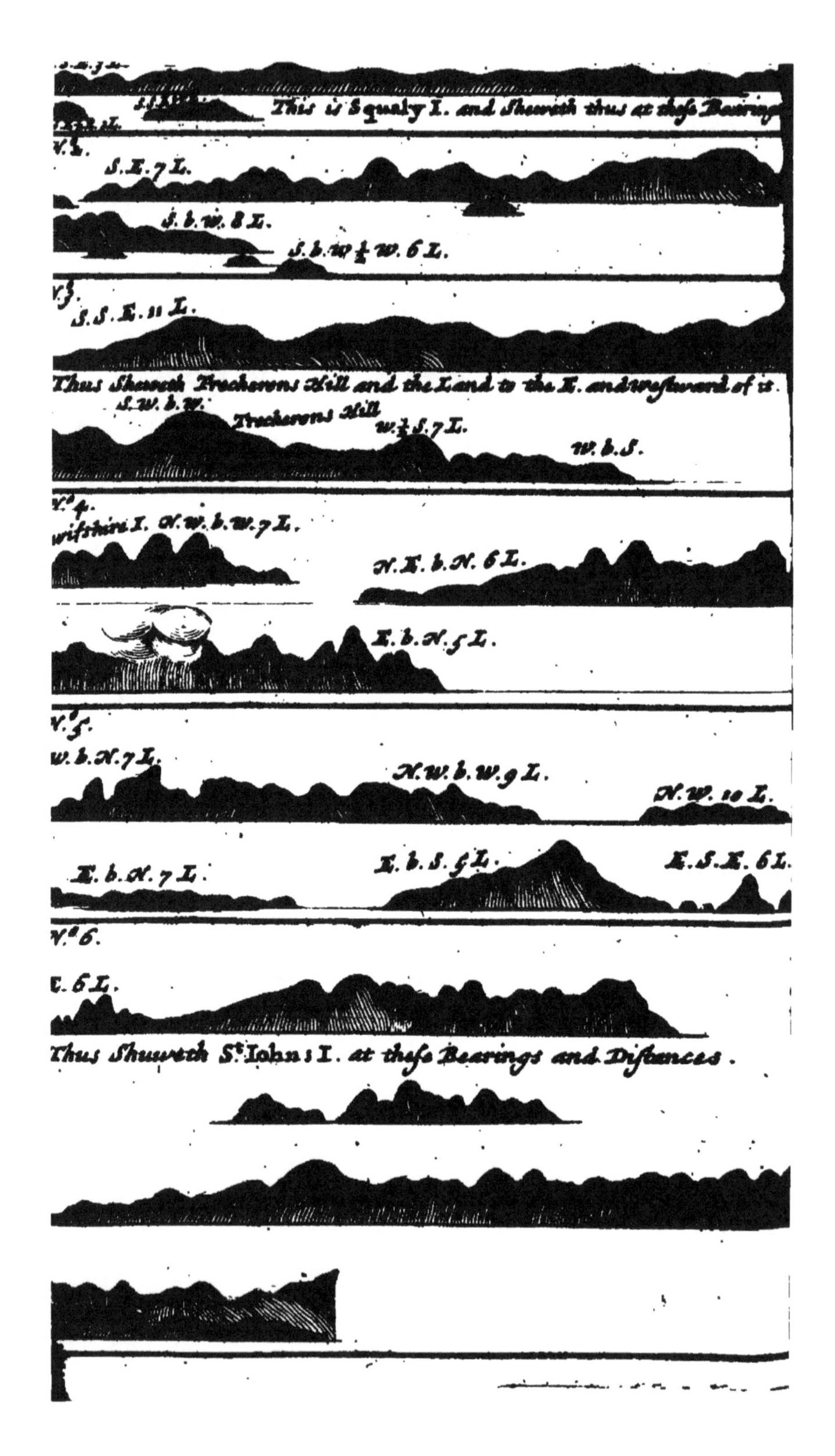

This is Squaly I. and Shewith thus at these Bearings
S.E.7 L.
S.b.W. 8 L.
S.b.W ½ W. 6 L.
N.3.
S.S.E. 11 L.
Thus Sheweth Trecherons Hill and the Land to the E. and westward of it.
S.W.b.W.
Trecherons Hill
W.½ S.7 L.
W.b.S.
N.o 4.
N.W.b.W.7 L.
N.E.b.N. 6 L.
E.b.N.5 L.
N.o 5.
W.b.N.7 L.
N.W.b.W.9 L.
N.W. 10 L.
E.b.N.7 L.
E.b.S.5 L.
E.S.E. 6 L.
N.o 6.
Thus Shuweth St. Iohns I. at these Bearings and Distances.

more Lightning than we had ſeen at any time this Voyage. This Morning we left a large high Iſland on our Larboard ſide, called in the *Dutch* Draughts *Wiſharts* Iſle, about ſix Leagues from the Main; and ſeeing many Smoaks upon the Main, I therefore ſteer'd towards it.

CHAP. IV.

The main Land of New Guinea. *Its Inhabitants.* Slingers *Bay. Small Islands.* Garret Dennis *Isle described. Its Inhabitants. Their Proes.* Anthony Caves *Island. Its Inhabitants. Trees full of Worms found in the Sea.* St. Johns *Island. The main Land of* New Guinea. *Its Inhabitants. The Coast described. Cape and Bay* St. George. *Cape* Orford. *Another Bay. The Inhabitants there. A large account of the Author's attempts to Trade with them. He names the place Port* Mountague. *The Country thereabouts described, and its produce. A Burning Island described. A new passage found.* Nova Britannia. Sir George Rooks *Island.* Long *Island, and* Crown *Island, discovered and described.* Sir R. Rich's *Island.*

Island. A Burning Island. A strange Spout. A Conjecture concerning a new passage Southward. King Williams *Island. Strange Whirlpools. Distance between Cape* Mabo, *and Cape* St. George, *computed.*

THE main Land, at this place, is high and mountainous, adorn'd with tall flourishing Trees; The sides of the Hills had many large Plantations and Patches of clear'd Land; which, together with the Smoaks we saw, were certain signs of its being well inhabited; and I was desirous to have some commerce with the Inhabitants. Being nigh the Shore, we saw first one Proe; a little after, two or three more; and at last a great many Boats came from all the adjacent Bays. VVhen they were forty six in Number, they approach'd so near us, that we could see each others signs, and hear each other speak; though we could not understand them, nor they us. They made signs for us to go in towards the Shore, pointing that way; it was squally VVeather, which at first made me cautious of going too near; but the Weather beginning to look pretty well, I endeavoured to get

An. 1699. into a Bay a-head of us, which we could have got into well enough at first; but while we lay by, we were driven so far to Leeward, that now it was more difficult to get in. The Natives lay in their Proes round us; to whom I shew'd Beads, Knives, Glasses, to allure them to come nearer; but they would not come so nigh, as to receive any thing from us. Therefore I threw out some things to them, *viz.* a Knife fastned to a piece of Board, and a Glass-bottle corked up with some Beads in it; which they took up and seemed well pleased. They often struck their left Breast with their right Hand, and as often held up a black Truncheon over their Heads, which we thought was a Token of Friendship; Wherefore we did the like. And when we stood in towards their Shore, they seem'd to rejoyce; but when we stood off, they frown'd, yet kept us Company in their Proes, still pointing to the Shore. About five a Clock we got within the Mouth of the Bay, and sounded several times, but had no Ground, though within a mile of the Shore. The Bason of this Bay was above two mile within us, into which we might have gone; but as I was not assured of Anchorage there, so I thought it not prudence to run in at this time; it being near Night, and seeing a black

Tor-

Tornado rising in the West, which I most fear'd: Besides, we had near two hundred Men in Proes close by us. And the Bays on the Shore were lined with Men from one end to the other, where there could not be less than three or four hundred more. What Weapons they had, we know not, nor yet their design. Therefore I had, at their first coming near us, got up all our small Arms, and made several put on Cartouch Boxes to prevent Treachery. At last I resolved to go out again: Which when the Natives in their Proes perceived, they began to fling Stones at us as fast as they could, being provided with Engines for that purpose; (wherefore I named this place *Slinger's Bay*:) But at the Firing of one Gun they were all amaz'd, drew off and flung no more Stones. They got together, as if consulting what to do; for they did not make in towards the Shore, but lay still, though some of them were killed or wounded; and many more of them had paid for their boldness, but that I was unwilling to cut off any of them; which if I had done, I could not hope afterwards to bring them to treat with me.

The next day we sailed close by an Island, where we saw many Smoaks, and Men in the Bays; out of which came two or three Canoas, taking much pains to

An. 1699. overtake us, but they could not, though we went with an easy Sail; and I could not now stay for them. As I past by the South-East point, I sounded several times within a mile of the Sandy Bays, but had no Ground: About three Leagues to the Northward of the South-East point, we opened a large deep Bay, secur'd from West-North-West and South-West Winds. There were two other Islands that lay to the North-East of it, which secur'd the Bay from North-East Winds; One was but small, yet woody; the other was a League long, inhabited and full of Coco-Nut-Trees: I endeavoured to get into this Bay; but there came such flaws off from the high Land over it, that I could not; Besides, we had many hard Squals, which deterr'd me from it; and Night coming on, I would not run any hazard; but bore away to the small inhabited Island, to see if we could get Anchoring on the East side of it. When we came there, we found the Island so narrow, that there could be no Shelter; therefore I tack'd and stood toward the greater Island again: And being more than midway between both, I lay by, designing to endeavour for Anchorage next Morning. Between seven and eight at Night, we spied a Canoa close by us; and seeing no more, suffered her to come aboard.

She

She had three Men in her, who brought off five Coco-nuts, for which I gave each of them a Knife and a ſtring of Beads, to encourage them to come off again in the Morning: But before theſe went away, we ſaw two more Canoas coming; therefore we ſtood away to the Northward from them, and then lay by again till Day. We ſaw no more Boats this Night; neither deſign'd to ſuffer any to come aboard in the dark.

An. 1699

By nine a Clock the next Morning, we were got within a League of the great Iſland, but were kept off by violent guſts of Wind. Theſe Squals gave us warning of their approach, by the Clouds which hung over the Mountains, and afterwards deſcended to the Foot of them; and then it is we expect them ſpeedily.

On the 3d of *March*, being about five Leagues to Leeward of the great Iſland, we ſaw the Main Land a-head; and another great high Iſland to Leeward of us, diſtance about ſeven Leagues; which we bore away for. It is called in the *Dutch* Draughts, *Garret Dennis* Iſle. It is about fourteen or fifteen Leagues round; high and mountainous, and very woody: Some Trees appeared very large and tall; and the Bays by the Sea-ſide are well ſtored vvith Coco-nut-Trees; vvhere vve alſo ſavv ſome ſmall Houſes. The ſides of

An. 1699. of the Mountains are thick ſet vvith Plantations; and the Mould in the new clear'd Land, ſeem'd to be of a brovvn reddiſh Colour. This Iſland is of no regular Figure, but is full of points ſhooting forth into the Sea; betvveen vvhich are many Sandy Bays, full of Coco-nut-Trees. The middle of the Iſle lies in 3 deg. 10 min. South Latitude. It is very populous: The Natives are very black, ſtrong, and vvell limb'd People; having great round Heads; their Hair naturally curl'd and ſhort, vvhich they ſhave into ſeveral forms, and dye it alſo of diverſe Colours, *viz.* Red, White and Yellovv. They have broad round Faces vvith great bottle Noſes, yet agreeable enough, till they disfigure them by Painting, and by wearing great things through their Noſes as big as a Mans Thumb and about four Inches long; theſe are run clear through both Noſtrils, one end coming out by one Cheek-Bone, and the other end againſt the other; and their Noſes ſo ſtretched, that only a ſmall ſlip of them appears about the Ornament. They have alſo great holes in their Ears, vvherein they vvear ſuch ſtuff as in their Noſes. They are very dextrous active Fellovvs in their Proes, vvhich are very ingeniouſly built. They are narrovv and long, vvith Outlagers on one ſide; the Head and Stern higher

higher than the rest, and carved into many Devices, *viz.* some Fowl, Fish, or a Mans Hand painted or carv'd: And though its but rudely done, yet the resemblance appears plainly, and shevvs an ingenious fancy. But vvith vvhat Instruments they make their Proes or carved Work, I knovv not; for they seem to be utterly ignorant of Iron. They have very neat Paddles, vvith vvhich they manage their Proes dextrously, and make great way through the Water. Their Weapons are chiefly Lances, Swords and Slings, and some Bows and Arrows: They have also Wooden Fissgigs, for striking Fish. Those that came to assault us in *Slingers* Bay on the Main, are in all respects like these; and I believe these are alike treacherous. Their Speech is clear and distinct; the words they used most, when near us, were *Vacousee Allamais*, and then they pointed to the Shore. Their signs of Friendship, are either a great Truncheon, or Bough of a Tree full of Leaves, put on their Heads; often striking their Heads with their Hands.

The next day, having a fresh Gale of Wind, we got under a high Island, about four or five Leagues round, very woody, and full of Plantations upon the sides of the Hills; and in the Bays, by the Water-side, are abundance of Coco-nut-Trees.

It

An. 1699. It lies in the Latitude of 3 deg. 25 min. South, and Meridian Distance from Cape *Mabo* 1316 m. On the South-East part of it are three or four other small woody Islands; one high and peek'd, the other low and flat; all bedeck'd with Coco-nut-Trees and other Wood. On the North there is another Island of an indifferent heighth, and of a somewhat larger circumference than the great high Island last mention'd. We past between this and the high Island. The high Island is called in the *Dutch* Draughts *Anthony Cave's Island*. As for the flat low Island, and the other small one, it is probable they were never seen by the *Dutch*; nor the Islands to the North of *Garret Dennis's Island*. As soon as we came near *Cave's Island*, some Canoas came about us, and made Signs for us to come ashore, as all the rest had done before; probably thinking we could run the Ship a-ground any where, as they did their Proes; for we saw neither Sail nor Anchor among any of them, though most *Eastern Indians* have both. These had Proes made of one Tree, well dug, with Outlagers on one side: They were but small, yet well shap'd. We endeavour'd to anchor, but found no Ground within a Mile of the Shore: We kept close along the North-side, still sounding till we came to the

North-

North-East end, but found no Ground; the Canoas still accompanying us; and the Bays were covered with Men going along as we sail'd: Many of them strove to swim off to us, but we left them astern. Being at the North-East point, we found a strong Current setting to the North-West; so that though we had steer'd to keep under the high Island, yet we were driven towards the flat one. At this time three of the Natives came aboard: I gave each of them a Knife, a Looking-Glass, and a String of Beads. I shew'd them Pumpkins and Coco-nut-shells, and made Signs to them to bring some aboard, and had presently three Coco-nuts out of one of the Canoas. I shewed them Nutmegs, and by their Signs I guess'd they had some on the Island. I also shew'd them some Gold-Dust, which they seem'd to know, and call'd out *Manneel*, *Manneel*, and pointed towards the Land. A while after these Men were gone, two or three Canoas came from the flat Island, and by Signs invited us to their Island; at which the others seem'd displeas'd, and us'd very menacing Gestures and (I believe) Speeches to each other. Night coming on, we stood off to Sea; and having but little Wind all Night, were driven away to the North-West. We saw many great Fires on the flat Island. These last Men

An. 1699. that came of to us, were all black, as those we had seen before, with frizled Hair: They were very tall, lusty, well-shap'd Men; They wear great things in their Noses, and paint as the others, but not much; They make the same Signs of Friendship, and their Language seems to be one: But the others had Proes, and these Canoas. On the sides of some of these, we saw the Figures of several Fish neatly cut; and these last were not so shy as the others.

Steering away from *Cave's Island* South-South-East, we found a strong Current against us, which set only in some places in Streams; and in them we saw many Trees and Logs of Wood, which drove by us. We had but little Wood aboard; wherefore I hoisted out the Pinnace, and sent her to take up some of this Drift-wood. In a little time she came aboard with a great Tree in a tow, which we could hardly hoist in with all our Tackles. We cut up the Tree and split it for Fire-wood. It was much worm-eaten, and had in it some live Worms above an Inch long, and about the bigness of a Goose-quill, and having their Heads crusted over with a thin Shell.

After this we passed by an Island, called by the *Dutch St John's Island*, leaving it to the North of us. It is about nine

or

or ten Leagues round, and very well adorn'd with lofty Trees. We saw many Plantations on the sides of the Hills, and abundance of Coco-nut-trees about them; as also thick Groves on the Bays by the Sea side. As we came near it, three Canoas came off to us, but would not come aboard. They were such as we had seen about the other Islands: They spoke the same Language, and made the same Signs of Peace; and their Canoas were such, as at *Cave's Island*.

An. 1699.

We stood along by *St John's Island*, till we came almost to the South-East Point; and then seeing no more Islands to the Eastward of us, nor any likelihood of anchoring under this, I steer'd away for the Main of *New-Guinea*; we being now (as I suppos'd) to the East of it, on this North side. My design of seeing these Islands as I past along, was to get wood and water; but could find no Anchor-Ground, and therefore could not do as I purpos'd. Besides, these Islands are all so populous, that I dar'd not send my Boat ashore, unless I could have anchor'd pretty nigh. Wherefore I rather chose to prosecute my Design on the Main, the Season of the Year being now at hand; for I judg'd the Westerly Winds were nigh spent.

On

An. 1699. On the 8th of *March*, we saw some Smoaks on the Main, being distant from it four or five Leagues. 'Tis very high, woody Land, with some spots of Savannah. About ten in the Morning six or seven Canoas came off to us: Most of them had no more than one Man in them; they were all black, with short curl'd Hair; having the same Ornaments in their Noses, and their Heads so shav'd and painted, and speaking the same words, as the Inhabitants of *Cave's* Island before-mentioned.

There was a Head-land to the Southward of us, beyond which seeing no Land, I supposed that from thence the Land trends away more Westerly. This Head-land lies in the Latitude of 5 deg. 2 min. South, and Meridian distance from Cape *Mabo*, one thousand two hundred and ninety Miles. In the Night we lay by, for fear of over-shooting this Headland. Between which and Cape St. *Maries*, the Land is high, Mountainous and VVoody; having many points of Land shooting out into the Sea, which make so many fine Bays. The Coast lies North-North-East and South-South-West.

The 9th in the Morning a huge black Man came off to us in a Canoa, but would not come aboard. He made the same signs of Friendship to us, as the rest we had

had met with; yet seem'd to differ in his Language, not using any of those words which the others did. VVe saw neither Smoaks nor Plantations near this Head-land. We found here Variation 1 deg. East.

In the Afternoon, as we plied near the Shore, three Canoas came off to us; one had four Men in her, the others two a-piece. That with the four Men, came pretty nigh us, and shew'd us a Coco-nut and Water in a Bamboo, making signs that there was enough ashore where they lived; they pointed to the place where they would have us go, and so went away. We saw a small round pretty high Island about a League to the North of this Head-land, within which there was a large deep Bay, whither the Canoas went; and we strove to get thither before Night, but could not; wherefore we stood off, and saw Land to the Westward of this Head-Land, bearing West by South half South, distance about ten Leagues; and, as we thought, still more Land bearing South-West by South, distance twelve or fourteen Leagues: But being clouded, it disappeared, and we thought we had been deceived. Before Night we opened the Head-Land fair, and I named it Cape *St. George*. The Land from hence trends away West-North-West about ten Leagues, which

An. 1699. which is as far as we could see it; and the Land that we saw to the Westward of it in the Evening, which bore West by South half South, was another point about ten Leagues from Cape *St. George*; between which there runs in a deep Bay for twenty Leagues or more. We saw some high Land in spots like Islands, down in that Bay at a great distance; but whether they are Islands, or the Main closing there, we know not. The next Morning we saw other Land to the South-East of the Westermost point, which till then was clouded; it was very high Land, and the same that we saw the day before, that disappear'd in a Cloud. This Cape *St. George* lies in the Latitude of 5 deg. 5 min. South; and Meridian distance from Cape *Mabo* a thousand two hundred and ninety Miles. The Island off this Cape, I called *St. Georges* Isle; and the Bay between it and the West-point, I named *St. Georges* Bay. *Note*, No *Dutch* Draughts go so far as this Cape, by ten Leagues. On the 10th in the Evening, we got within a League of the Westermost Land seen, which is pretty high and very woody, but no appearance of Anchoring. I stood off again, designing (if possible) to ply to and fro in this Bay, till I found a conveniency to Wood and Water. We saw no more Plantations, nor Coco-nut-Trees;

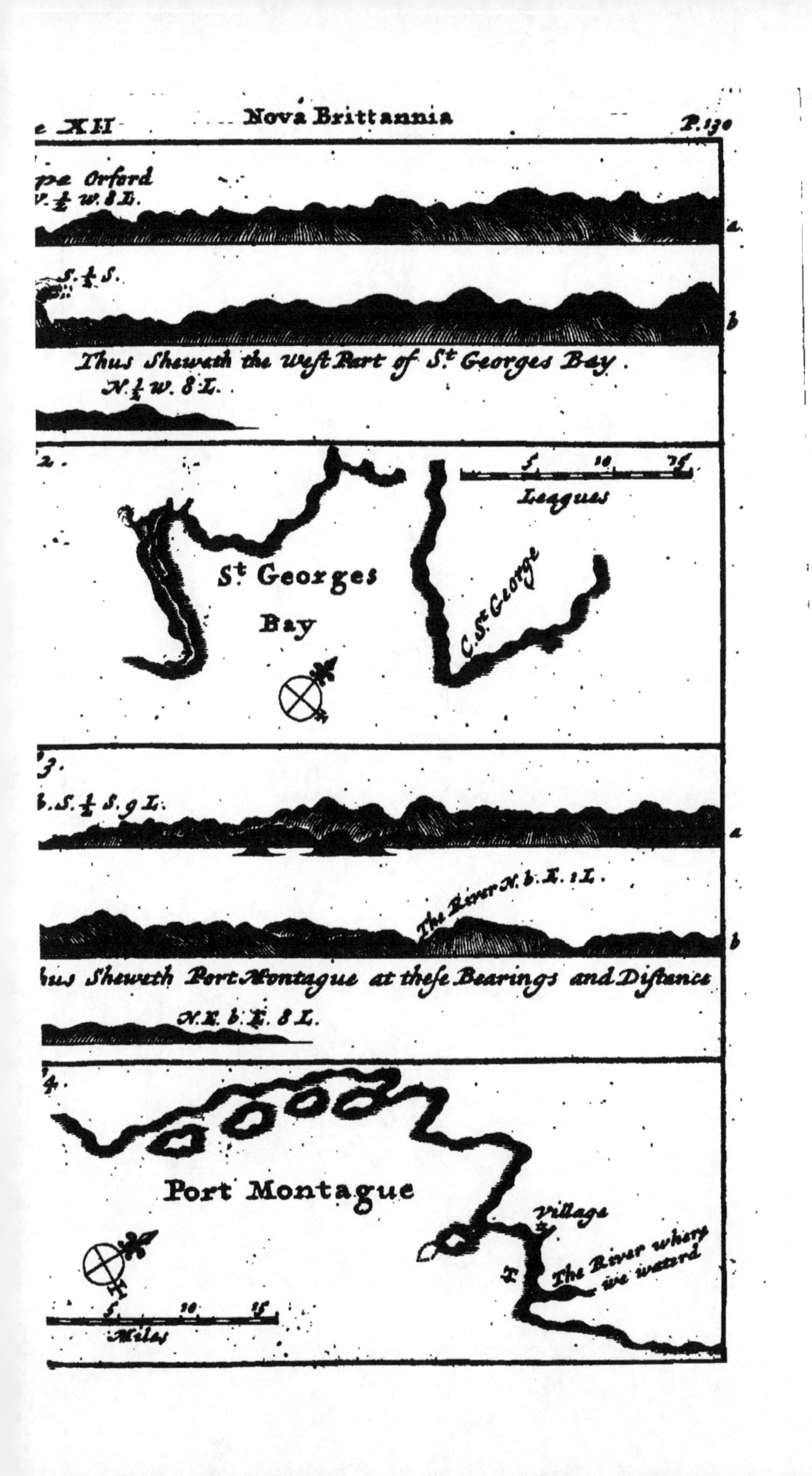

Nova Brittannia
P.130
pe Orford
Thus Sheweth the West Part of St Georges Bay .
N. ½ W. 8 L.
Leagues
St Georges
Bay
C. S^t. George
The River N. b. E. 1 L.
Thus Sheweth Port Montague at these Bearings and Distance
N. E. b. E. 8 L.
Port Montague
Village
The River where we watterd
Miles

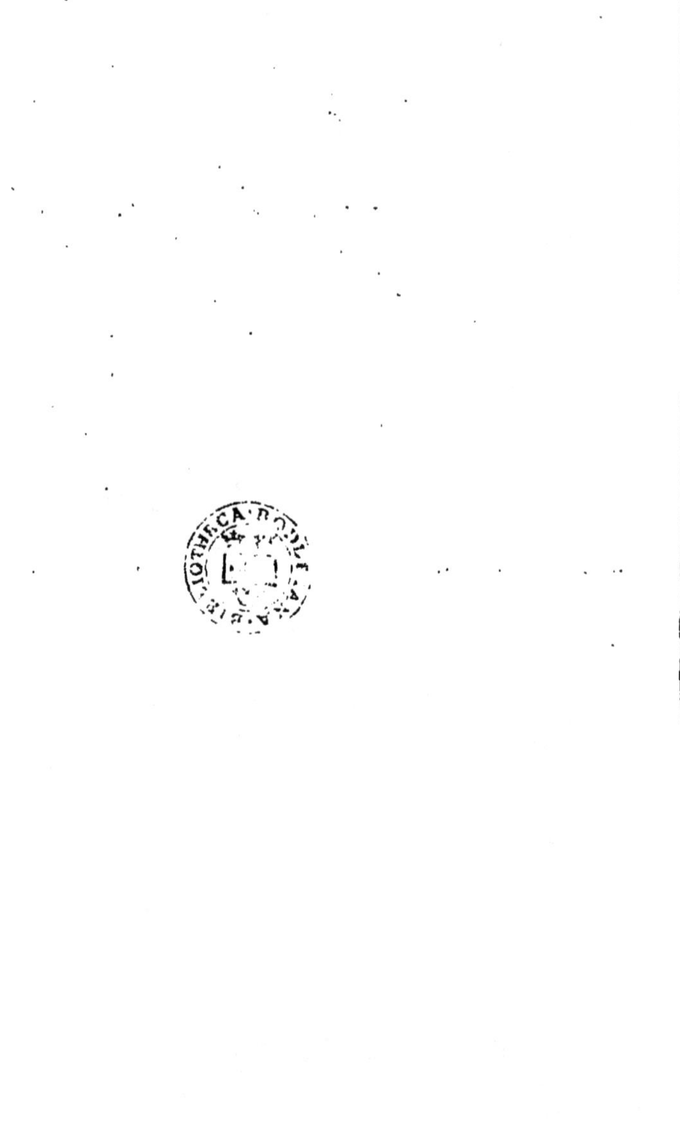

Trees; yet in the Night we discerned a small Fire right against us. The next Morning we saw a Burning Mountain in the Country. It was round, high, and peaked at top (as most *Vulcano's* are,) and sent forth a great quantity of Smoak. We took up a Log of drift Wood, and split it for Firing; in which we found some small Fish.

An. 1699

The day after, we past by the South-West Cape of this Bay, leaving it to the North of us: When we were abreast of it, I called my Officers together, and named it Cape *Orford*, in honour of my noble Patron; drinking his Lordship's health. This Cape bears from Cape *St. George* South-West about eighteen Leagues. Between them there is a Bay about twenty five Leagues deep, having pretty high Land all round it, especially near the Capes, though they themselves are not high. Cape *Orford* lies in the Latitude of 5 deg. 24 min. South, by my Observation; and Meridian distance from Cape *St. George*, forty four miles West. The Land trends from this Cape North-West by West into the Bay, and on the other side South-West *per Compass*, which is South-West 9 deg. West, allowing the Variation which is here 9 deg. East. The Land on each side of the Cape, is more Savannah than wood Land; and is highest on the North-

An. 1699 North-West side. The Cape it self is a Bluff-point, of an indifferent heighth, with a flat Table Land at top. When we were to the South-West of the Cape, it appeared to be a low point shooting out; which you cannot see when abreast of it. This Morning we struck a Log of Drift-wood with our Turtle-Irons, hoisted it in and split it for Fire-wood. Afterwards we struck another, but could not get it in. There were many Fish about it.

We steer'd along South-West as the Land lies, keeping about six Leagues off the Shore; and being desirous to cut Wood and fill VVater, if I saw any conveniency, I lay by in the Night, because I would not miss any place proper for those ends, for fear of wanting such Necessaries as we could not live without. This Coast is high and mountainous, and not so thick with Trees as that on the other side of Cape *Orford*.

On the 14th, seeing a pretty deep Bay a-head, and some Islands where I thought we might ride secure, we ran in towards the Shore, and saw some Smoaks. At ten a Clock we saw a point, which shot out pretty well into the Sea, with a Bay within it, which promised fair for VVater; and we stood in, with a moderate Gale. Being got into the Bay within the Point,

Point, we ſaw many Coco-nut-Trees, Plantations, and Houſes. VVhen I came within four or five mile of the Shore, ſix ſmall Boats came off to view us, with about forty Men in them all. Perceiving that they only came to view us, and would not come aboard, I made ſigns and waved to them to go aſhore; but they did not or would not underſtand me; therefore I whiſtled a ſhot over their Heads out of my Fowling-piece, and then they pull'd away for the Shore as hard as they could. Theſe were no ſooner aſhore, but we ſaw three Boats coming from the Iſlands to Leeward of us, and they ſoon came within call; for we lay becalm'd. One of the Boats had about forty Men in her, and was a large well built Boat; the other two, were but ſmall. Not long after, I ſaw another Boat coming out of that Bay where I intended to go: She likewiſe was a large Boat, with a high Head and Stern Painted, and full of Men; this I thought came off to fight us, as 'tis probable they all did; therefore I fired another ſmall ſhot over the great Boat that was nigh us, which made them leave their babling and take to their Paddles. VVe ſtill lay becalm'd; and therefore they rowing wide of us, directed their courſe toward the other great Boat that was coming off: VVhen

An. 1699. they were pretty near each other, I caus'd the Guuner to fire a Gun between them, which he did very dextrously; it was loaden with round and Partridge shot; the last dropt in the VVater somewhat short of them, but the round shot went between both Boats, and grazed about a hundred yards beyond them; this so affrighted them, that they both rowed away for the Shore as fast as they could, without coming near each other; and the little Boats made the best of their way after them: And now having a gentle Breeze at South-South-East, we bore into the Bay after them. VVhen we came by the point, I saw a great number of Men peeping from under the Rocks: I ordered a shot to be fired close by, to scare them. The shot graz'd between us and the point; and mounting again, flew over the point, and graz'd a second time just by them. VVe were obliged to sail along close by the Bays; and seeing multitudes setting under the Trees, I ordered a third Gun to be Fired among the Coco-nut-Trees, to scare them; for my business being to VVood and VVater, I thought it necessary to strike some terrour into the Inhabitants, who were very numerous, and (by what I saw now, and had formerly experienced,) treacherous. After this I sent my Boat to sound; they had

had first forty, then thirty, and at last twenty Fathom VVater. VVe followed the Boat, and came to anchor about a quarter of a mile from the Shore, in twenty six Fathom VVater, fine black Sand and Oaze. VVe rode right against the Mouth of a small River, where I hoped to find fresh VVater. Some of the Natives standing on a small point at the Rivers Mouth, I sent a small shot over their Heads to fright them; which it did effectually. In the Afternoon I sent my Boat ashore to the Natives who stood upon the point by the Rivers Mouth with a present of Coco-nuts; when the Boat was come near the Shore, they came running into the VVater, and put their Nuts into the Boat. Then I made a signal for the Boat to come aboard, and sent both it and the Yawle into the River to look for fresh VVater, ordering the Pinnace to lye near the Rivers Mouth, while the Yawle went up to search. In an hours time they return'd aboard with some Barreccoes full of fresh Water, which they had taken up about half a mile up the River. After which, I sent them again with Casks; ordering one of them to fill Water, and the other to watch the motion of the Natives, least they should make any opposition; but they did not, and so the Boats return'd a little before Sun-

An. 1699. ſet with a Tun and half of Water; and the next day by noon brought aboard about ſix Tun of Water.

I ſent aſhore Commodities to purchaſe Hogs, *&c.* being informed that the Natives have plenty of them, as alſo of Yamms and other good Roots; But my Men returned without getting any thing that I ſent them for; the Natives being unwilling to Trade with us: Yet they admir'd our Hatchets and Axes; but would part with nothing but Coco-nuts; which they us'd to climb the Trees for; and ſo ſoon as they gave them our Men, they beckon'd to them to be gone; for they were much afraid of us.

The 18th, I ſent both Boats again for Water, and before noon they had filled all my Casks. In the Afternoon I ſent them both to cut Wood; but ſeeing about forty Natives ſtanding on the Bay at a ſmall diſtance from our Men, I made a ſignal for them to come aboard again; which they did, and brought me word that the Men which we ſaw on the Bay were paſſing that way, but were afraid to come nigh them. At four a Clock I ſent both the Boats again for more Wood, and they return'd in the Evening. Then I called my Officers to conſult whether it were convenient to ſtay here longer, and endeavour a better acquaintance with theſe

people;

people; or go to Sea. My design of tarrying here longer, was, if possible, to get some Hogs, Goats, Yamms or other Roots; as also to get some knowledge of the Country and its product. My Officers unanimously gave their opinions for staying longer here. So the next day I sent both Boats ashore again, to fish, and to cut more VVood. VVhile they were ashore, about thirty or forty Men and Women past by them; they were a little afraid of our People at first; but upon their making signs of Friendship, they past by quietly; the Men finely bedeck'd with Feathers of divers Colours about their Heads, and Lances in their Hands; the VVomen had no Ornament about them, nor any thing to cover their Nakedness, but a bunch of small green Boughs, before and behind, stuck under a string which came round their Wastes. They carried large Baskets on their Heads, full of Yamms. And this I have observ'd amongst all the wild Natives I have known, that they make their Women carry the burdens, while the Men walk before, without any other load than their Arms and Ornaments. At noon our Men came aboard with the Wood they had cut, and had catch'd but six Fishes at four or five hauls of the Sain, though we saw abundance of Fish leaping in the Bay all the day long.

In

An. 1699. In the Afternoon I sent the Boats ashore for more Wood; and some of our Men went to the Natives Houses, and found they were now more shy than they us'd to be; had taken down all the Coco-nuts from the Trees, and driven away their Hogs. Our People made signs to them to know what was become of their Hogs, &c. The Natives pointing to some Houses in the bottom of the Bay, and imitating the noise of those Creatures, seem'd to intimate that there were both Hogs and Goats of several sizes, which they express'd by holding their Hands abroad at several distances from the Ground.

At night our Boats came aboard with Wood; and the next Morning I went my self with both Boats up the River to the Watering-place, carrying with me all such Trifles and Iron-work as I thought most proper to induce them to a Commerce with us; but I found them very shy and roguish. I saw but two Men and a Boy: One of the Men by some signs was perswaded to come to the Boat's side, where I was; to him I gave a Knife, a String of Beads, and a Glass-bottle; the Fellow call'd out, *Cocos, Cocos*, pointing to a Village hard by, and signified to us that he would go for some; but he never return'd to us. And thus they had frequently,

quently of late ſerved our Men. I took eight or nine Men with me, and march-ed to their Houſes, which I found very mean; and their Doors made faſt with Withes.

I viſited three of their Villages; and finding all the Houſes thus abandon'd by the Inhabitants, who carried with them all their Hogs, *&c*, I brought out of their Houſes ſome ſmall Fiſhing-nets in recompence for thoſe things they had re-ceiv'd of us. As we were coming away, we ſaw two of the Natives; I ſhewed them the things that we carried with us, and called to them, *Cocos*, *Cocos*, to let them know that I took theſe things be-cauſe they had not made good what they had promis'd by their Signs, and by their calling out *Cocos*. While I was thus em-ploy'd, the Men in the Yawle filled two Hogsheads of Water, and all the Barre-coes. About one in the afternoon I came aboard, and found all my Officers and Men very importunate to go to that Bay where the Hogs were ſaid to be. I was loath to yield to it, fearing they would deal too roughly with the Natives. By two a-Clock in the afternoon many black Clouds gather'd over the Land, which I thought would deter them from their En-terprize; but they ſolicited me the more to let them go. At laſt I conſented, ſend-ing

An. 1699. ing those Commodities I had ashore with me in the Morning, and giving them a strict charge to deal by fair means, and to act cautiously for their own Security. The Bay I sent them to, was about two Miles from the Ship. Assoon as they were gone, I got all things ready, that, if I saw occasion, I might assist them with my great Guns. When they came to land, the Natives in great Companies stood to resist them; shaking their Lances, and threatning them; And some were so daring, as to wade into the Sea, holding a Target in one Hand and a Lance in the other. Our Men held up to them such Commodities as I had sent, and made signs of Friendship; but to no purpose; for the Natives waved them off. Seeing therefore they could not be prevailed upon to a friendly Commerce, my Men, being resolved to have some Provision among them, fired some Muskets to scare them away; which had the desired effect upon all but two or three, who stood still in a menacing posture, till the boldest dropt his Target and ran away; They suppos'd he was shot in the Arm: He and some others felt the smart of our Bullets, but none were kill'd; our design being rather to fright than to kill them. Our Men landed, and found abundance of tame Hogs running among the Houses.

They ſhot down nine, which they brought away, beſides many that ran away wounded. They had but little time; for in leſs than an hour after they went from the Ship, it began to rain: Wherefore they got what they could into the Boats; for I had charg'd them to come away if it rain'd. By that time the Boat was aboard, and the Hogs taken in, it clear'd up; and my Men deſir'd to make another trip thither before night; This was about five in the Evening; and I conſented, giving them order to repair on Board before night. In the cloſe of the Evening they returned accordingly, with eight Hogs more, and a little live Pig; and by this time the other Hogs were jerk'd and ſalted. Theſe that came laſt, we only dreſt and corn'd till morning; and then ſent both Boats aſhore for more Refreſhments, either of Hogs or Roots: But in the night the Natives had convey'd away their Proviſions of all ſorts. Many of them were now about the Houſes, and none offer'd to reſiſt our Boats landing, but on the contrary were ſo amicable, that one Man brought ten or twelve Coco-nuts, left them on the Shore after he had ſhew'd them to our Men, and went out of ſight. Our People finding nothing but Nets and Images, brought ſome of them away; which two of my Men brought aboard in a

An. 1699.

ſmall

An. 1699. ſmall Canoa; and preſently after, my Boats came off. I order'd the Boatſwain to take care of the Nets, till we came at ſome place where they might be diſpoſed of for ſome Refreſhment for the uſe of all the Company: The Images I took into my own cuſtody.

In the Afternoon I ſent the Canoa to the place from whence ſhe had been brought; and in her, two Axes, two Hatchets (one of them helv'd,) ſix Knives, ſix Looking-glaſſes, a large bunch of Beads, and four Glaſs-bottles Our Men drew the Canoa aſhore, placed the things to the beſt advantage in her, and came off in the Pinnace which I ſent to guard them. And now being well ſtock'd with Wood, and all my Water-casks full, I reſolv'd to ſail the next Morning. All the time of our ſtay here, we had very fair Weather; only ſometimes in the Afternoon we had a Shower of Rain, which laſted not above an hour at moſt: Alſo ſome Thunder and Lightning, with very little VVind. VVe had Sea and Land-breezes; the former between the South-South-Eaſt, and the latter from North-Eaſt to North-Weſt.

This place I named Port *Mountague*, in honour of my noble Patron. It lies in the Latitude of 6 deg. 10 min. South, and Meridian diſtance from Cape *St. George*,

An. 1699. 6.10 S 2.31 W

George, one hundred fifty one miles West. The Country hereabouts is Mountainous and Woody, full of rich Valleys and pleasant fresh Water-brooks. The Mould in the Valleys is deep and yellowish; that on the sides of the Hills of a very brown Colour, and not very deep, but rocky underneath; yet excellent planting Land. The Trees in general are neither very streight, thick, nor tall; yet appear green and pleasant enough: Some of them bore Flowers, some Berries, and others big Fruits; but all unknown to any of us. Coco-nut-Trees thrive very well here; as well on the Bays by the Sea-side, as more remote among the Plantations. The Nuts are of an indifferent size, the Milk and Kernel very thick and pleasant. Here is Ginger, Yamms, and other very good Roots for the Pot; that our Men saw and tasted. What other Fruits or Roots the Country affords, I know not. Here are Hogs and Dogs; other Land-Animals we saw none. The Fowls we saw and knew, were Pidgeons, Parrots, Cockadores and Crows like those in *England*; a sort of Birds about the bigness of a Black-Bird, and smaller Birds many. The Sea and Rivers have plenty of Fish; we saw abundance, though we catch'd but few, and these were Cavallies, Yellow-tails and Whip-rays.

We

An. 1699.

We departed from hence on the 22d of *March*, and on the 24th in the Evening we ſaw ſome high Land bearing North-Weſt half Weſt; to the Weſt of which we could ſee no Land, though there appeared ſomething like Land bearing Weſt a little Southerly; but not being ſure of it, I ſteered Weſt-North-Weſt all Night, and kept going on with an eaſie Sail, intending to coaſt along the Shore at a diſtance. At ten a Clock I ſaw a great Fire bearing North-Weſt by VVeſt, blazing up in a Pillar, ſometimes very high for three or four Minutes, then falling quite down for an equal ſpace of time; ſometimes hardly viſible, till it blazed up again. I had laid me down, having been indiſpoſed this three days: But upon a ſight of this, my chief Mate called me; I got up and view'd it for about half an Hour, and knew it to be a burning Hill by its intervals: I charg'd them to look well out, having bright Moon-light. In the Morning I found that the Fire we had ſeen the Night before, was a burning Iſland; and ſteer'd for it. We ſaw many other Iſlands, one large high Iſland, and another ſmaller, but pretty high. I ſtood near the *Vulcano*, and many ſmall low Iſlands with ſome Shoals.

March the 25th 1700, in the Evening we came within three Leagues of this Burning-hill, being at the same time two Leagues from the Main. I found a good Channel to pass between them, and kept nearer the Main than the Island. At seven in the Evening I sounded, and had fifty two Fathom fine Sand and Oaze. I stood to the Northward to get clear of this Streight, having but little VVind and fair VVeather. The Island all Night vomited Fire and Smoak very amazingly; and at every Belch we heard a dreadful Noise like Thunder, and saw a flame of Fire after it, the most terrifying that ever I saw. The intervals between its Belches, were about half a minute; some more, others less: Neither were these Pulses or Eruptions alike; for some were but faint Convulsions, in comparison of the more vigorous; yet even the weakest vented a great deal of Fire; but the largest made a roaring Noise, and sent up a large Flame 20 or 30 yards high; and then might be seen a great stream of Fire running down to the Foot of the Island, even to the Shore. From the Furrows made by this descending Fire, we could in the day time see great Smoaks arise, which probably were made by the Sulphureous Matter thrown out of the Funnel at the top, which tumbling

An. 1700. down to the bottom, and there lying in a heap, burn'd till either consumed or extinguished; and as long as it burn'd and kept its heat, so long the Smoak ascended from it; which we perceived to increase or decrease, according to the quantity of Matter discharged from the Funnel. But the next Night, being shot to the Westward of the Burning-Island, and the Funnel of it lying on the South side, we could not discern the Fire there, as we did the Smoak in the day when we were to the Southward of it. This Vulcano lies in the Latitude of 5 deg. 33 min. South, and Meridian distance from Cape St. *George*, three hundred thirty two miles West.

The Eastermost part of *New Guinea* lies forty miles to the Westward of this Tract of Land; and by Hydrographers they are made joyning together: But here I found an opening and passage between, with many Islands; the largest of which, lye on the North side of this Passage or Streight. The Channel is very good, between the Islands and the Land to the Eastward. The East part of *New Guinea*, is high and mountainous, ending on the North-East with a large Promontory, which I nam'd *King William's* Cape, in honour of his present Majesty. We saw some Smoaks on it; and leaving

Table XIII Dampiers Passage and Islands on ye Coast of N. Guinea,

No. 1.

S.W. ½ W. 9 L.

W. 12 L.

Thus shews ye S.W. Land when your in ye S. Part of ye Entrance of Capt. Damp...

N.W. b. W. ½ W. 8 L.

W. b. N. ½ N. 5 L.

N.W. b. N. 9 L.

N.N.W. ½ W.

N. ½ W. 7 L.

N.E. b. E. 10 L.

No. 2.

E. b. N. 5 L.

S. ½ E. 5 L.

S.W. b. W. ½ W. 3 L.

S.W. b. S. 6 L.

W. b. S. 5 L.

W. 2 L.

No. 3.

S.W. b. W. 3 L.

W. b. S. ½ S. 5 L.

S.S.W. ½ W. 6 L.

W. ½ S. 2 L.

No. 4.

N.N.W. 4 L.

W. b. S. 11 L.

W

No. 5.

S.E. ½ E. 6 L.

S. b. W. 6 L.

S.W. b. W. 3 L.

W. 2 ½ L.

eaving it on our Larboard side, steer'd away near the East Land; which ends with two Remarkable Capes or Heads, distant from each other about six or seven Leagues. Within each Head were two very remarkable Mountains, ascending very gradually from the Sea side; which afforded a very pleasant and agreeable Prospect. The Mountains and lower Land were pleasantly mixt with VVood-Land and Savannahs. The Trees appeared very Green and Flourishing; and the Savannahs seem'd to be very smooth and even; No Meadow in *England* appears more Green in the Spring, than these. We saw Smoaks, but did not strive to Anchor here; but rather chose to get under one of the Islands, (where I thought I should find few or no Inhabitants,) that I might repair my Pinnace; which was so crazy that I could not venture ashore any where with her. As we stood over to the Islands, we look'd out very well to the North, but could see no Land that way; by which I was well assur'd that we were got through, and that this East Land does not joyn to *New Guinea*; Therefore I named it *Nova Britannia*. The North-VVest Cape, I called Cape *Glocester*, and the South-VVest point Cape *Ann*; and the North-

An. 1700.

An. 1700. VVest Mountain, which is very remarkable, I call'd Mount *Glocester.*

This Island which I called *Nova Britannia*, has about 4 deg. of Latitude: The Body of it lying in 4 deg. and the Northermost part in 2 deg. 30 min. and the Southermost in 6 deg. 30 min. South. It has about 5 deg. 18 min. Longitude from East to West. It is generally high, mountainous Land, mixt with large Valleys; which, as well as the Mountains, appeared very Fertile; and in most places that we saw, the Trees are very large, tall and thick. It is also very well inhabited with strong well-limb'd *Negroes*, whom we found very daring and bold at several Places. As to the product of it, I know no more than what I have said in my Account of *Port Mountague*: But it is very probable this Island may afford as many rich Commodities as any in the World; and the Natives may be easily brought to Commerce, though I could not pretend to it under my present Circumstances.

Being near the Island to the Northward of the *Vulcano*, I sent my Boat to sound, thinking to Anchor here; but she return'd and brought me word that they had no ground, till they met with a Riff of Coral Rocks about a mile from the Shore. Then I bore away to the North

North side of the Island, where we found no Anchoring neither. We saw several People, and some Coco-nut-Trees; but could not send ashore for want of my Pinnace which was out of order. In the Evening I stood off to Sea, to be at such a distance, that I might not be driven by any Current upon the Shoals of this Island, if it should prove calm. We had but little Wind, especially the beginning of the Night; But in the Morning I found my self so far to the West of the Island, that, the Wind being at East-South-East, I could not fetch it; Wherefore I kept on to the Southward, and stemm'd with the Body of a high Island about eleven or twelve Leagues long, lying to the Southward of that which I before designed for. I named this Island Sir *George Rook's* Island.

We also saw some other Islands to the Westward; which may be better seen in my Draught of these Lands, than here described. But seeing a very small Island lying to the North-West of the long Island which was before us, and not far from it; I steer'd away for that; hoping to find Anchoring there: And having but little Wind, I sent my Boat before to sound; which, when we were about two miles distance from the Shore, came on board and brought me word that there

An. 1700. was good Anchoring in thirty or forty Fathom Water, a mile from the Isle, and within a riff of the Rocks which lay in a half Moon, reaching from the North part of the Island to the South-East; So at noon we got in and anchored in thirty-six Fathom, a Mile from the Isle.

In the Afternoon I sent my Boat ashore to the Island, to see what convenience there was to haul our Vessel ashore in order to be mended, and whither we could catch any Fish. My Men in the Boat rowed about the Island, but could not Land by reason of the Rocks and a great Surge running in upon the Shore. We found Variation here, 8 deg. 25 min. West.

I design'd to have stay'd among these Islands till I had got my Pinnace refitted; but having no more than One Man who had skill to work upon her, I saw she would be a long time in repairing; (which was one great Reason why I could not prosecute my Discoveries further:) And the Easterly Winds being set in, I found I should scarce be able to hold my Ground.

The 31st in the Forenoon we shot in between two Islands, lying about four Leagues asunder; with intention to pass between them. The Southermost is a long Island, with a high Hill at each

end;

end; this I named *Long Island*. The Northermost is a round high Island, towering up with several Heads or Tops, something resembling a Crown; this I named *Crown-Isle*, from its form. Both these Islands appear'd very pleasant, having spots of green Savannahs mixt among the Wood-land: The Trees appeared very Green and Flourishing, and some of them looked white and full of Blossoms. We past close by *Crown-Isle*; saw many Coco-nut-Trees on the Bays and the sides of the Hills; and one Boat was coming off from the Shore, but return'd again. We saw no Smoaks on either of the Islands, neither did we see any Plantations; and it is probable they are not very well peopled. We saw many Shoals near *Crown-Island*, and riffs of Rocks running off from the Points, a mile or more into the Sea. My Boat was once over-board, with design to have sent her ashore; but having little Wind, and seeing some Shoals, I hoisted her in again, and stood off out of danger.

In the Afternoon, seeing an Island bearing North-West by West, we steer'd away North-West by North, to be to the Northward of it. The next Morning, being about mid-way from the Islands we left yesterday, and having this to the

An. 1700. Westward of us; the Land of the Main of New Guinea within us to the Southward, appear'd very high. When we came within four or five Leagues of this Island to the West of us, four Boats came off to view us; one came within call, but return'd with the other three without speaking to us: So we kept on for the Island; which I named Sir *R. Rich*'s Island. It was pretty high, woody, and mixt with Savannah's like those formerly mentioned. Being to the North of it, we saw an opening between it and another Island two Leagues to the West of it, which before appear'd all in One. The Main seemed to be high Land, trending to the Westward.

On *Tuesday* the 2d of *April*, about eight in the Morning, we discovered a high peeked Island to the Westward, which seem'd to smoak at its top. The next day we past by the North side of the Burning Island, and saw a Smoak again at its top; but the vent lying on the South side of the Peek, we could not observe it distinctly, nor see the Fire. We afterwards opened three more Islands, and some Land to the Southward, which we could not well tell whether it were Islands or part of the Main. These Islands are all high, full of fair Trees and spots of green Savannahs; as well the

Burn-

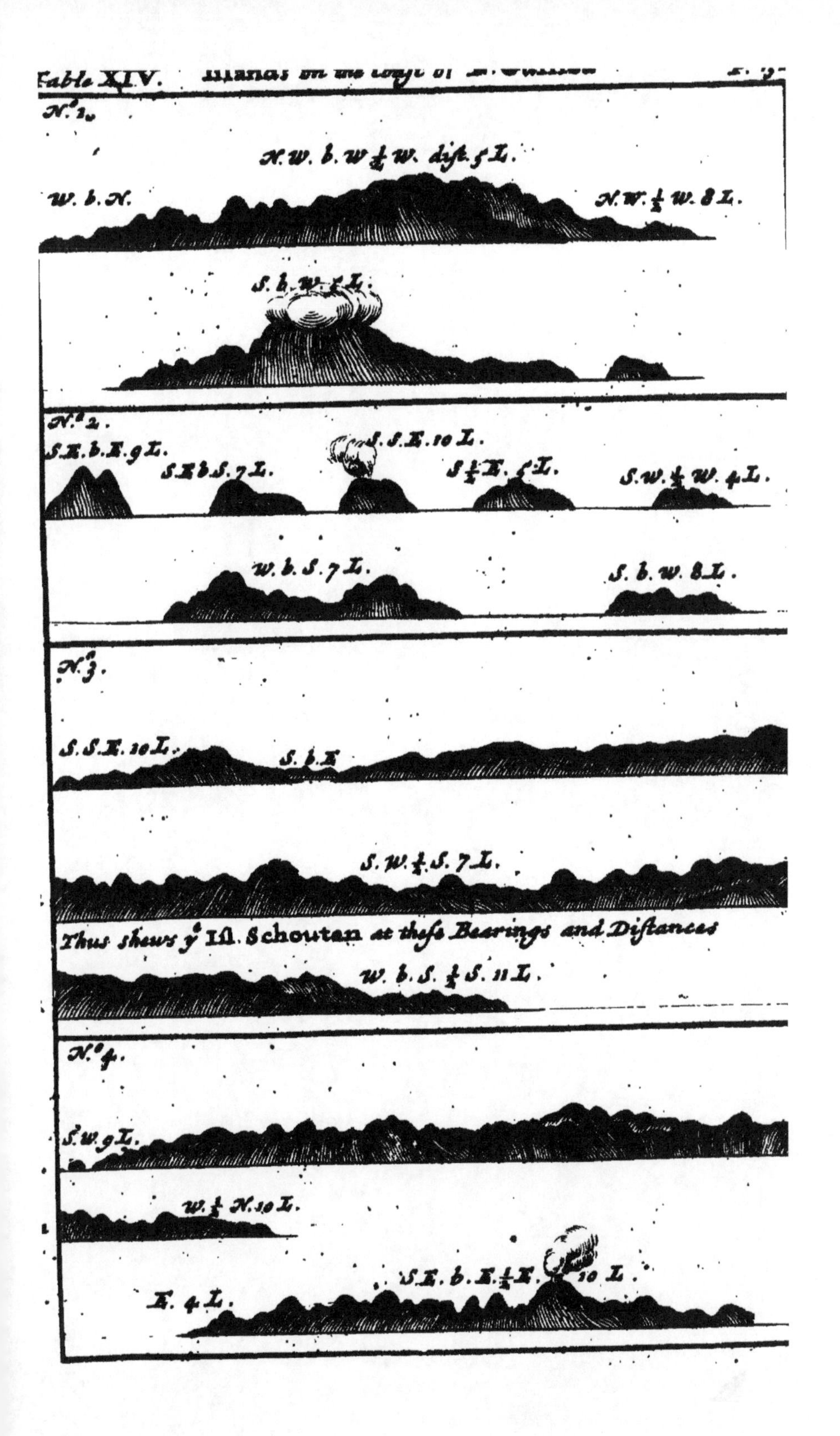
Table XIV. Islands on the coast of N. Guinea
N.º 1.
N.W. b. W ½ W. dist. 5 L.
W. b. N.
N.W. ½ W. 8 L.
S. b. W. 5 L.
N.º 2.
S.E. b. E. 9 L.
S.E b. S. 7 L.
S. S. E. 10 L.
S ½ E. 5 L.
S.W. ½ W. 4 L.
W. b. S. 7 L.
S. b. W. 8 L.
N.º 3.
S. S. E. 10 L.
S. b. E
S. W. ½ S. 7 L.
Thus shews yᵉ Ifl. Schouten at these Bearings and Distances
W. b. S. ½ S. 11 L.
N.º 4.
S. W. 9 L.
W. ½ N. 10 L.
S.E. b. E. ½ E. 10 L.
E. 4 L.

Burning Isle as the rest; but the Burning Isle was more round and peek'd at top, very fine Land near the Sea, and for two thirds up it. We also saw another Isle sending forth a great Smoak at once; but it soon vanished, and we saw it no more. We saw also among these Islands three small Vessels with Sails, which the people on *Nova Britannia* seem wholly ignorant of.

An. 1700.

The 11th at noon, having a very good observation, I found my self to the Northward of my reckoning; and thence concluded that we had a Current setting North-West, or rather more Westerly, as the Land lies. From that time to the next Morning, we had fair clear Weather, and a fine moderate Gale from South-East to East by North: But at day break, the Clouds began to fly, and it Lightned very much in the East, South-East, and North-East. At Sun-rising, the Sky look'd very Red in the East near the Horizon; and there were many black Clouds both to the South and North of it. About a quarter of an hour after the Sun was up, there was a Squall to the Windward of us; when on a sudden one of our Men on the Fore-castle called out that he saw something a-stern, but could not tell what: I look'd out for it, and immediately saw a Spout beginning to work

An. 1700. work within quarter of a mile of us, exactly in the Wind. We presently put right before it. It came very swiftly, whirling the Water up in a Pillar about six or seven yards high. As yet I could not see any Pendulous Cloud, from whence it might come; and was in hopes it would soon lose its force. In four or five minutes time it came within a Cables length of us, and past away to Leeward; and then I saw a long pale Stream, coming down to the whirling Water. This Stream was about the bigness of a Rainbow: The upper end seem'd vastly high, not descending from any dark Cloud, and therefore the more strange to me; I never having seen the like before. It past about a mile to Leeward of us, and then broke. This was but a small Spout, not strong nor lasting; yet I perceived much Wind in it, as it past by us. The Current still continued at North-West a little Westerly, which I allow'd to run a mile *per* hour.

By an observation the 13th at noon, I found my self 25 min. to the Northward of my reckoning; whether occasion'd by bad Steerage, a bad Account, or a Current, I could not determine: But was apt to judge it might be a complication of all; for I could not think it was wholly the Current, the Land here lying East

East by South, and West by North, or a little more Northerly and Southerly. We had kept so nigh as to see it, and at farthest had not been above twenty Leagues from it, but sometimes much nearer; and it is not probable that any Current should set directly off from a Land. A Tide indeed may; but then the Flood has the same force to strike in upon the Shore, as the Ebb to strike off from it: But a Current must have set nearly along Shore, either Easterly or Westerly; and if any thing Northerly or Southerly, it could be but very little in comparison of its East or West course, on a Coast lying as this doth; Which yet we did not perceive. If therefore we were deceiv'd by a Current, it is very probable that the Land is here disjoyn'd, and that there is a passage through to the Southward, and that the Land from *King William's* Cape to this place is an Island, separated from *New Guinea* by some Streight, as *Nova Brittannia* is by that which we came through. But this being at best but a probable conjecture, I shall insist no farther upon it.

The 14th we passed by *Schouten's* Island and *Providence* Island, and found still a very strong Current setting to the North-West. On the 17th we saw a high Mountain on the Main, that sent forth great

An. 1700. great quantities of Smoak from its top: This *Vulcano* we did not see in our Voyage out. In the Afternoon we discovered *King William's* Island, and crowded all the Sail we could, to get near it before Night; thinking to lye to the Eastward of it till day, for fear of some Shoals that lye at the West-end of it. Before Night we got within two Leagues of it, and having a fine Gale of Wind and a light Moon, I resolv'd to pass through in the Night; which I hop'd to do before twelve a-clock, if the Gale continued; but when we came within 2 miles of it, it fell calm; yet afterwards by the help of the Current, a small Gale, and our Boat, we got through before day. In the Night we had a very fragrant smell from the Island. By Morning-light we were got two Leagues to the Westward of it; and then were becalm'd all the Morning; and met such whirling Tides, that when we came into them, the Ship turn'd quite round; and though sometimes we had a small Gale of Wind, yet she could not feel the Helm when she came into these Whirlpools: Neither could we get from amongst them, till a brisk Gale sprung up; yet we drove not much any way, but whirl'd round like a Top. And those Whirlpools were not constant to one place, but drove about strangely; and

some-

ſometimes we ſaw among them large riplings of the Water; like great Overfalls, making a fearful Noiſe. I ſent my Boat to ſound, but found no Ground.

The 18th, Cape *Mabo* bore S. diſtance nine Leagues. By which account it lies in the Latitude of 50 min. South, and Meridian diſtance from Cape St. *George* one thouſand two hundred forty three miles. St. *Johns* Iſle lies forty eight miles to the Eaſt of Cape St. *George*; which being added to the diſtance between Cape St. *George* and Cape *Mabo*, makes one thouſand two hundred ninety one Meridional parts; which was the furtheſt that I was to the Eaſt. In my outward bound Voyage I made Meridian diſtance between Cape *Mabo* and Cape St. *George*, one thouſand two hundred and ninety miles; and now in my return, but one thouſand two hundred forty three; which is forty ſeven ſhort of my diſtance going out. This difference may probably be occaſion'd by the ſtrong Weſtern Current which we found in our return, which I allowed for after I perceived it; and though we did not diſcern any Current when we went to the Eaſtward, except when near the Iſlands; yet it is probable we had one againſt us, though we did not take notice

An. 1700. tice of it becauſe of the ſtrong Weſterly Winds. *King Willam's* Iſland lies in the Latitude of 21 min. South, and may be ſeen diſtinctly off of Cape *Mabo*.

In the Evening we paſt by Cape *Mabo*; and afterwards ſteer'd away South-Eaſt half Eaſt, keeping along the Shore, which here trends South-Eaſterly. The next Morning, ſeeing a large opening in the Land, with an Iſland near the South ſide; I ſtood in, thinking to Anchor there. When we were ſhot in within two Leagnes of the Iſland, the Wind came to the Weſt, which blows right into the Opening. I ſtood in to the North Shore; intending, when I came pretty nigh, to ſend my Boat into the Opening, and ſound, before I would adventure in: We found ſeveral deep Bays, but no Soundings within two miles of the Shore; therefore I ſtood off again. Then ſeeing a ripling under our Lee, I ſent my Boat to ſound on it; which return'd in half an hour, and brought me word that the ripling we ſaw was only a Tide, and that they had no Ground there.

CHAP.

CHAP. V.

The A.'s return from the Coast of New Guinea. *A deep Channel. Strange Tides. The Island* Ceram *described. Strange Fowls. The Islands* Bonao, Bouro, Misacombi, Pentare, Laubana, *and* Potoro. *The Passage between* Pentare *and* Laubana. *The Island* Timor. Babao *Bay. The Island* Rotte. *More Islands than are commonly laid down in the Draughts. Great Currents. Whales. Coast of* New Holland. *The* Tryal-Rocks. *The Coast of* Java. Princes Isle. *Streights of* Sunda. Thwart-the-way *Island. Indian Proes, and their Traffick. Passage through the Streight. Arrival at* Batavia.

THE Wind seeming to incline to East, as might be expected according to the Season of the Year; I rather choes

An. 1700. chose to shape my Course as these Winds would best permit, than strive to return the same way we came; which, for many Leagues, must have been against this Monsoon: Though indeed on the other hand, the dangers in that way, we already knew; but what might be in this, by which we now proposed to return, we could not tell.

We were now in a Channel about eight or nine Leagues wide, having a range of Islands on the North side, and another on the South side, and very deep Water between, so that we had no Ground. The 22d of *April* in the Morning, I sent my Boat ashore to an Island on the North side, and stood that way with the Ship. They found no Ground till within a Cables length of the Shore, and then had Coral Rocks; so that they could not catch any Fish, though they saw a great many. They brought aboard a small Canoa, which they found a-drift. They met with no Game ashore, save only one party-colour'd Parrakite. The Land is of an indifferent height; very Rocky, yet cloathed with tall Trees, whose bare Roots run along upon the Rocks. Our People saw a Pond of Salt Water, but found no fresh. Near this Island we met a pretty strong Tide, but found neither Tide nor Current off at some distance. On

On the 24th, being about two Leagues from an Iſland to the Southward of us, we came over a Shoal on which we had but five Fathom and a half. We did not deſcrie it, till we ſaw the Ground under us. In leſs than half an hour before, the Boat had been ſounding in diſcoloured Water, but had no Ground. VVe mann'd the Boat preſently, and tow'd the Ship about; and then ſounding, had twelve, fifteen and ſeventeen Fathom, and then no Ground with our Hand-lead. The Shoal was rocky; but in twelve and fifteen Fathom, we had oazy Ground.

An. 1700.

We found here very ſtrange Tides, that ran in Streams, making a great Sea; and roaring ſo loud, that we could hear them before they came within a mile of us. The Sea round about them ſeem'd all broken, and toſſed the Ship ſo that ſhe would not anſwer her Helm. Theſe riplings commonly laſted ten or twelve minutes, and then the Sea became as ſtill and ſmooth as a Mill-pond. VVe ſounded often when in the midſt of them, and afterwards in the ſmooth VVater; but found no Ground, neither could we perceive that they drove us any way.

VVe had in one Night ſeveral of theſe Tides, that came moſt of them from the

An. 1700. VVest; and the VVind being from t quarter, we commonly heard them a lo time before they came; and sometin lowered our Top-sails, thinking it w a gust of Wind. They were of gre length from North to South, but th breadth not exceeding two hundre yards, and they drove a great pace For though we had little VVind to mov us, yet these would soon pass away and leave the VVater very smooth; an just before we encountred them, we me a great swell, but it did not break.

The 26th, we saw the Island *Ceram*; and still met some riplings, but much fainter than those we had the two preceedings days. VVe sail'd along the Island *Ceram* to the VVestward, edging in withal, to see if peradventure we might find a Harbour to Anchor in, where we might water, trim the Ship, and refresh our Men.

In the Morning we saw a Sail to the North of us, steering in for the VVest-end of *Ceram*, as we likewise were. In the Evening, being near the Shore on the North-side of the Island, I stood off to Sea with an easy Sail; intending to stand in for the Shore in the Morning, and try to find Anchoring, to fill VVater, and get a little Fish for refreshment. Accordingly in the Morning early, I stood

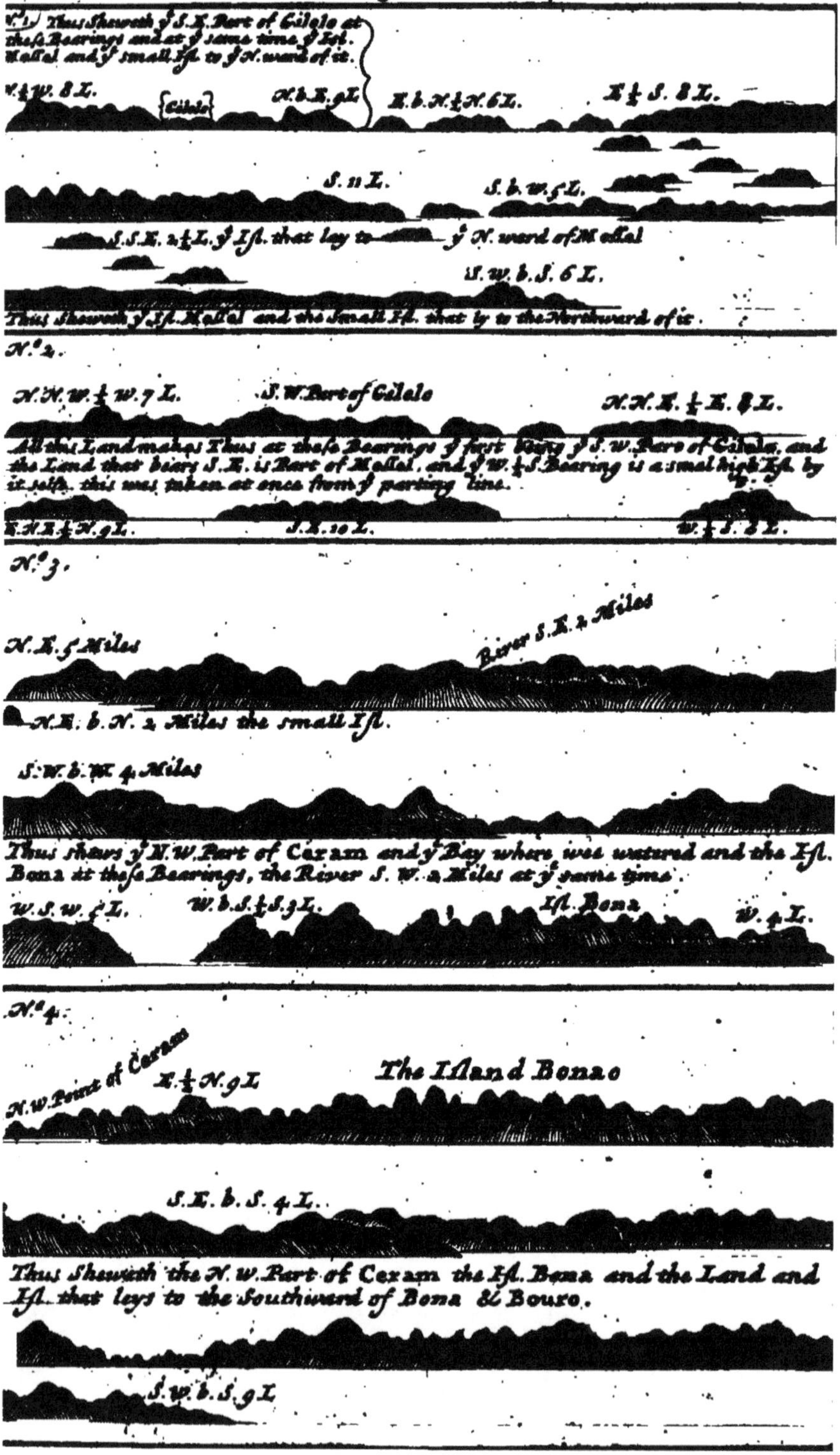

N°1. Thus Sheweth ye S.E. Part of Gilolo at these Bearings and at ye same time ye Isl. Mallal and ye small Isl. to ye N.ward of it.

N.½W. 8 L. Gilolo N.b.E. 9 L. E.b.N.½N. 6 L. E½S. 8 L.

S. 11 L. S.b.W. 5 L.

S.S.E. 2½ L. ye Isl. that lay to ye N.ward of Mallal

S.W.b.S. 6 L.

Thus Sheweth ye Isl. Mallal and the Small Isl. that ly to the Northward of it.

N°2.

N.N.W.½W. 7 L. S.W. Part of Gilolo N.N.E.½E. 8 L.

All this Land makes Thus at these Bearings ye first being ye S.W. Part of Gilolo, and the Land that bears S.E. is Part of Mallal, and ye W.½S. Bearing is a small high Isl. by it selfe. this was taken at once from ye parting line.

E.N.E.½N. 9 L. S.E. 10 L. W.½S. 8 L.

N°3.

N.E. 5 Miles River S.E. 2 Miles

N.E.b.N. 2 Miles the small Isl.

S.W.b.W. 4 Miles

Thus shews ye N.W. Part of Ceram and ye Bay where wee watered and the Isl. Bona at these Bearings, the River S.W. 2 Miles at ye same time.

W.S.W. 5 L. W.b.S.½S. 3 L. Isl. Bona W. 4 L.

N°4.

N.W. Point of Ceram E.½N. 9 L. The Island Bonao

S.E.b.S. 4 L.

Thus Sheweth the N.W. Part of Ceram the Isl. Bona and the Land and Isl. that leys to the Southward of Bona & Bouro.

S.W.b.S. 9 L.

ſtood in with the North-VVeſt point of *Ceram*; leaving a ſmall Iſland, called *Bonao*, to the VVeſt. The Sail we ſaw the day before, was now come pretty nigh us, ſteering in alſo (as we did) between *Ceram* and *Bonao*. I ſhortned Sail a little for him; and when he got a-breaſt of us, not above two miles off, I ſent my Boat aboard. It was a *Dutch* Sloop, come from *Terranate*, and bound for *Amboyna*: My Men whom I ſent in the Boat, bought five Bags of new Rice, each containing about one hundred and thirty pounds, for ſix *Spaniſh* Dollars. The Sloop had many rare Parrots aboard for Sale, which did not want price. A *Malayan* Merchant aboard, told our Men, that about ſix Months ago he was at *Bencola*, and at that time the Governour either dyed or was kill'd, and that the Commander of an *Engliſh* Ship then in that Road ſucceeded to that Government.

In the Afternoon, having a Breeze at North and North-North-Eaſt, I ſent my Boat to ſound, and ſtanding after her with the Ship, anchored in thirty Fathom VVater oazy Sand, half a mile from the Shore, right againſt a ſmall River of freſh Water. The next Morning I ſent both the Boats aſhore to fiſh; they return'd about ten a Clock, with a

An. 1700 few Mullets and three or four Cavallies, and ſome Pan-Fiſh. We found Variation here, 2 deg. 15 min. Eaſt.

When the Sea was ſmooth by the Land-winds, we ſent our Boats aſhore for Water; who, in a few turns, filled all our Casks.

The Land here is low, ſwampy and woody; the Mould is a dark Gray, friable Earth. Two Rivers came out within a Bow-ſhot of each other, juſt oppoſite to the place where we rode: One comes right down out of the Country; and the other from the South, running along by the Shore, not Muſquet ſhot from the Sea-ſide. The Northermoſt River is biggeſt, and out of it we filled our Water; our Boats went in and out at any time of Tide. In ſome places the Land is overflown with freſh Water, at full Sea. The Land hereabouts is full of Trees unknown to us, but none of them very large or high; the Woods yield many wild Fruits and Berries, ſuch as I never ſaw elſewhere. We met with no Land-Animals. The Fowls we found, were Pidgeons, Parrots, Cockadores, and a great number of ſmall Birds unknown to me. One of the Maſter's Mates killed two Fowls as big as Crows; of a black Colour, excepting that the Tails were all white. Their Necks were pretty long, one

This Bird was taken on the Coast of New Guinea

A Stately Land Fowl found on the Coast of New Guinea described Page 93

A Strange Land Fowl found on the Island Ceram. described Page 165.

Page 165

one of which was of a Saffron-colour, the other black. They had very large Bills, much like a Rams-horn; their Legs were ſtrong and ſhort, and their claws like a Pidgeons; their Wings of an ordinary length: Yet they make a great noiſe when they fly, which they do very heavily. They feed on Berries, and perch on the higheſt Trees. Their Fleſh is ſweet; I ſaw ſome of the ſame Species at *New Guinea*, but no where elſe.

May the 3d, at ſix in the Morning we weigh'd, intending to paſs between *Bonao* and *Ceram*; but preſently after we got under Sail, we ſaw a pretty large Proe coming about the North-Weſt point of *Ceram*. Wherefore I ſtood to the North to ſpeak with her, putting aboard our Enſign. She ſeeing us coming that way, went into a ſmall Creek, and skulked behind a point a while: At laſt diſcovering her again, I ſent my Boat to ſpeak with her; but the Proe row'd away, and would not come nigh it. After this, finding I could not paſs between *Bonao* and *Ceram*, as I purpoſed; I ſteer'd away to the North of it.

This *Bonao* is a ſmall Iſland, lying about four Leagues from the North-Weſt point of *Ceram*. I was inform'd by the *Dutch* Sloop before-mentioned, that notwithſtanding its ſmallneſs, it hath one

 fine

An. 1700. fine River, and that the *Dutch* are there ſettled. Whether there be any Natives on it, or not, I know not; nor what its produce is. They further ſaid, that the *Ceramers* were their mortal Enemies; yet that they were ſettled on the Weſtermoſt point of *Ceram*, in ſpite of the Natives.

The next day, as we approach'd the Iſland *Bouro*, there came off from it a very fragrant ſcent, much like that from *King William's* Iſland; and we found ſo ſtrong a Current ſetting to the Weſtward, that we could ſcarce ſtem it. We plied to get to the Southward, intending to paſs between *Bouro* and *Keelang*.

In the Evening, being near the Weſt-end of *Bouro*, we ſaw a Brigantine to the North-Weſt of us, on the North-ſide of *Bouro*, ſtanding to the Eaſtward. I would not ſtand Eaſt or Weſt for fear of coming nigh the Land which was on each ſide of us, *viz.* *Bouro* on the Weſt and *Keelang* on the Eaſt. The next Morning we found our ſelves in Mid-channel between both Iſlands; and having the Wind at South-Weſt we ſteer'd South-South-Eaſt, which is right through between both. At eleven a Clock it fell calm, and ſo continued till noon; by that time the Brigantine, which we ſaw a-Stern the Night before, was got two

or

N. 1.

This Isl. makes Thus at these Bearings when ye Isl. Bona shews at ye other side

S. W. ½ S. 11 L. S. W. b. W. 14. L.

N.° 2.

S. W. ½ S. 5 L. W. b. S. 5 L.

W. ½ S. 12 L.

Thus Shews the Isl. Ambolow and Bouro at these Bearings

N. N. W. 7 L.

N.° 3.

E. S. E. ½ S. 10 L. Ambo S. E. 9 L.

S. S. E. 7 L. S. ½ E. 8 L.

At these Bearings Sheweth ye Isl. Ambo and ye Islands as you see to ye South Westward of it.

S. S. W. ½ W. 9 L. S. W. 11. L. S. W. ½ W. 12. L.

N.° 4.

W. ½ N. 11 L.

The Passage wch wee came through

Thus Sheweth ye Islands Laubana and Panterra at these Bearings wch wee came between at ye Bearings N. W. b. N. also ye Islands between that and Ambo as you see.

Part of Ambo.

An. 1700.

or three Leagues a-head of us. It is probable she met a strong Land-wind in the Evening, which continued all Night; she keeping nearer the Shore, than I could safely do. She might likewise have a Tide or Current setting Easterly, where she was; though we had a Tide setting Northwardly against us, we being in Mid-channel.

About eight at Night, the Brigantine which we saw in the day, came close along by us on our Weather-side: Our Guns were all ready before Night, Matches lighted, and small Arms on the Quarter-Deck ready loaden. She standing one way, and we another; we soon got further asunder. But I kept good watch all the Night, and in the Morning saw her a-Stern of us, standing as we did. At ten a Clock, having little Wind, I sent the Yawle aboard of her. She was a Chinese Vessel, laden with Rice, Arrack, Tea, Porcellane, and other Commodities, bound for *Amboyna*. The Commander said that his Boat was gone ashore for Water, and ask'd our Men if they saw her; for she had been wanting two or three days, and they knew not what was become of her. They had their Wives and Children aboard, and probably came to settle at some new *Dutch* Factory. The Commander also

An. 1700 inform'd us, that the *Dutch* had lately settled at *Ampulo*, *Menippe*, *Bonao*, and on a point of *Ceram*. The next day we past out to the Southward between *Keelang* and *Bouro*. After this, we had for several days a Current setting Southerly, and a great tumbling Sea, occasion'd more by the strong Current than by Winds, as was apparent by the jumping of its Waves against each other; and by Observation I found twenty-five miles more Southing then our Course gave us.

On the 14th we discovered the Island *Misacomby*, and the next day sail'd along to the West on the North side of the Island. In some Charts it is called *Omba*; it is a mountainous Island, spotted with Woods and Savannahs; about twenty Leagues long, and five or six broad. We saw no signs of Inhabitants on it. We fell in nearest to the West end of it; and therefore I chose to pass on to the Westward, intending to get through to the Southward between this and the next Isle to the West of it, or between any other two Islands to the West, where I should meet with the clearest passage; because the Winds were now at North-East and East-North-East, and the Isle lies nearly East and West; so that if the Winds continued, I might be a

long

long time in getting to the East end of it, which yet I knew to be the best passage. In the Night, being at the West-end, and seeing no clear passage, I stood off with an easie Sail, and in the Morning had a fine Land-wind, which would have carried us five or six Leagues to the East, if we had made the best of it; but we kept on only with a gentle Gale, for fear of a Westerly Current. In the Morning, finding we had not met with any Current as we expected; assoon as it was Light, we made Sail to the Westward again

After noon, being near the end of the Isle *Pentare*, which lies West from *Missacomby*, we saw many Houses and Plantations in the Country, and many Coco-nut-Trees growing by the Sea side. We also saw several Boats sailing cross a Bay or Channel at the West end of *Missacomby*, between it and *Pentare* We had but little Wind, and that at North, which blows right in, with a swell rowling in withal; wherefore I was afraid to venture in, though probably there might be good Anchoring, and a Commerce with the Natives. I continued steering to the VVest, because the Night before, at Sun-setting, I saw a small round high Island to the West of *Pentare*, where I expected a good passage.

An. 1700. We could not that day reach the West end of *Pentare*, but saw a deep Bay to the West of us, where I thought might be a passage through, between *Pentare* and *Laubana*. But as yet the Lands were shut one within an other, that we could not see any passage. Therefore I ordered to sail seven Leagues more Westerly, and lye by till next day. In the Morning we look'd out for an Opening, but could see none; yet by the distance and bearing of a high round Island called *Potoro*, we were got to the West of the Opening, but not far from it. Wherefore I tack'd and stood to the East; and the rather, because I had reason to suppose this to be the passage we came through in the *Cygnet* mentioned in my Voyage round the World; but I was not yet sure of it, because we had rainy Weather, so that we could not now see the Land so well as we did then. We then accidentally saw the Opening, at our first falling in with the Islands; which now was a work of some time and difficulty to discover. However before ten a Clock we saw the Opening plain; and I was the more confirm'd in my knowledge of this passage, by a Spit of Sand and two Islands at the North-East part of its entrance. The Wind was at South-South-West, and we plied

to

to get through before Night; for we found a good Tide helping us to the South. About seven or eight Leagues to the West of us we saw a high round piked Mountain, from whose top a Smoak seem'd to ascend as from a *Vulcano*. There were three other very high piked Mountains, two on the East, and one on the West of that which smoak-ed.

In our plying to get through between *Pentare* and *Laubana*, we had (as I said) a good Tide or Current setting us to the Southward. And it is to be observed, that near the Shores in these parts we commonly find a Tide setting Northwardly or Southwardly, as the Land lyes; but the Northwardly Tide sets not above three hours in twelve, having little strength; and sometimes it only checks the contrary Current, which runs with great violence, especially in narrow passes, such as this, between two Islands. It was twelve at Night before we got clear of two other small Islands, that lay on the South side of the passage; and there we had a very violent Tide setting us through against a brisk Gale of Wind. Notwithstanding which, I kept the Pinnace out, for fear we should be becalm'd. For this is the same place, through which I passed in the Year one thousand six

hundred

An. 1700. hundred eighty seven, mentioned in my Voyage round the World, (*pag.* 459.) Only then we came out between the Western small Island and *Laubana*, and now we came through between the two small Islands. We sounded frequently, but had no Ground. I said there, that we came through between *Omba* and *Pentare*: For we did not then see the Opening between those two Islands; which made me take the West side of *Pentare* for the West end of *Omba*, and *Laubana* for *Pentare*. But now we saw the Opening between *Omba* and *Pentare*; which was so narrow, that I would not venture through: Besides, I had now discovered my mistake, and hop'd to meet with the other passage again, as indeed we did, and found it to be bold from side to side, which in the former Voyage I did not know. After we were through, we made the best of our way to *Timor*; and on *May* the 18th in the Morning, we saw it plain, and made the high Land over *Laphao* the *Portugueze* Factory, as also the high Peak over our first Watering-place, and a small round Island about mid-way between them.

We coasted along the Island *Timor*, intending to touch at *Babao*, to get a little Water and Refreshments. I would not go into the Bay where we first water'd, ter'd, because of the Currents which there whirle about very strangely, especially at Spring-tides, which were now setting in; besides, the South-East Winds come down in flaws from the Mountains, so that it would have been very dangerous for us. Wherefore we crowded all the Sail we could, to get to *Babao* before Night, or at least to get sight of the Sandy Island at the entrance of the Bay; but could not. So we plied all Night; and the next Morning entered the Bay.

An. 1700.

There being good Ground all over this Bay, we anchored at two a Clock in thirty Fathom Water, soft oazy Ground. And the Morning after I sent my Boat ashore with the Sain to Fish. At noon she return'd and brought enough for all the Ship's Company. They saw an *Indian* Boat at a round Rocky Island about a mile from them.

On the 22d, I sent my Boat ashore again to Fish: At noon she return'd with a few Fish, which serv'd me and my Officers. They catch'd one Whiteing, the first I had seen in these Seas. Our people went over to the Rocky Island, and there found several Jarrs of Turtle, and some hanging up a drying, and some Cloaths; their Boat was about a mile off, striking Turtle. Our Men left all as they found.

An. 1700. found. In the Afternoon, a very large Shark came under our Stern; I never had ſeen any near ſo big before. I put a piece of Meat on a Hook for him, but he went a-Stern and return'd no more. About Mid-night, the Wind being pretty moderate, I weigh'd and ſtood into the bottom of the Bay, and ran over nearer the South Shore, where I thought to lye and water, and at convenient times get Fiſh for our refreſhment. The next Morning, I ſent my Pinnace with two Hogſheads and ten Barreccoes for Water; They return'd at noon with the Casks full of Water, very thick and muddy, but ſweet and good. VVe found Variation, 15 min. VVeſt.

This Afternoon, finding that the Breezes were ſet in here, and that it blew ſo hard that I could neither fiſh nor fill Water without much difficulty and hazard of the Boat; I reſolved to be gone, having good quantity of VVater aboard. Accordingly at half an hour after two in the Morning we weighed with the Wind at Eaſt by South, and ſtood to Sea. We coaſted along by the Iſland *Rotte*, which is high Land, ſpotted with VVoods and Savannahs. The Trees appear'd ſmall and ſhrubby, and the Savannahs dry and ruſty. All the North-ſide, has Sandy Bays by the Sea. We ſaw no Houſes nor Plantations. The

The next day we crowded all the Sail we could, to get to the West of all the Isles before Night, but could not; for at six in the Evening we saw Land bearing South-VVest by VVest. For here are more Islands than are laid down in any Draughts that I have seen. Wherefore I was oblig'd to make a more Westerly Course than I intended, till I judg'd we might be clear of the Land. And when we were so, I could easily perceive by the Ships motion. For till then, being under the Lee of the Shore, we had smooth Water; but now we had a troubled Sea which made us dance lustily. This turbulent Sea, was occasion'd in part by the Current; which setting out slanting against the Wind, was by it raised into short cockling Seas. I did indeed expect a South-West Current here, but not so very strong as we found it.

On the 26th we continued to have a very strong Current setting Southwardly; but on what point exactly, I know not. Our whole distance by Log was but eighty two miles, and our difference of Latitude since Yesterday-noon by observation one hundred miles, which is eighteen miles more than the vvhole distance; and our course, allovving no Lee-vvay at all, vvas South 17 deg. West, vvhich gives but seventy six miles difference of Latitude,

An. 1700. Latitude, tvventy four leſs than we found by obſervation. I did expect (as has been ſaid) vve might meet a great Current ſetting to the South yeſterday, becauſe there is a conſtant Current ſetting out from among thoſe Iſlands vve paſs'd through betvveen *Timor* and the Isles to the Weſt of it, and, 'tis probable, in all the other Openings betvveen the Islands, even from the Eaſt end of *Java* to the end of all that Range that runs from thence, both to the Eaſt and Weſt of *Timor*: But being got ſo far out to Sea as we were, though there may be a very great Current, yet it does not ſeem probable to me that it ſhould be of ſo great ſtrength as we now found: For both Currents and Tides looſe their force in the open Sea, where they have room to ſpread; and it is only in narrow places, or near Head-lands, that their force is chiefly felt. Beſides in my opinion, it ſhould here rather ſet to the VVeſt than South; being open to the narrow Sea, that divides *New-Holland* from the range of Islands before-mentioned.

The 27th, we found that in the laſt twenty four hours vve had gone nine miles leſs South than the Log gave: So that 'tis probable vve vvere then out of the Southern Current, vvhich vve felt

ſo

so much before. We savv many Tropick-Birds about us. And found Variation 1 deg. 25 min. West.

On *June* the 1st, we saw several Whales, the first we had at this time seen on the Coast; But when we were here before, we saw many; at which time we were nearer the Shore than now. The Variation now, was 5 deg. 38 min. West.

I design'd to have made *New Holland* in about the Latitude of 20 deg. and steer'd Courses by day to make it, but in the Night could not be so bold; especially since we had sounding. This Afternoon I steer'd in South-VVest, till six a Clock; then it blowing fresh, and Night coming on, I steer'd West-South-West, till we had forty Fathom; and then stood West, which course carries along Shore. In the Morning again from six to twelve I steer'd West-South-West, to have made the Land; but, not seeing it, I judged we were to the West of it. Here is very good Soundings on this Coast. When we past this way to the Eastward, we had, near this Latitude of 19 deg. 50 min. thirty-eight Fathom, about eighteen Leagues from the Land: But, this time, we saw not the Land. The next Morning I saw a great many Scuttle-Fish-bones, which was a sign that we

An. 1700. were not far from the Land. Alſo a great many Weeds continually floating by us.

VVe found the Variation increaſe conſiderably as we went VVeſtward. For on the 3d, it was 6 deg. 10 min. Weſt; on the 4th, 6 deg. 20 min. and on the 6th, 7 deg. 20 min. That Evening we ſaw ſome Fowls like *Men of War Birds* flying North-Eaſt, as I was told; for I did not ſee them, having been indiſpoſed theſe three or four days.

On the 11th we found the Variation 8 deg. 1 min. Weſt; on the 12th, 6 deg. 0 min. I kept on my Courſe to the Weſtward till the 15th, and then altered it. My deſign was to ſeek for the *Tryal Rocks*; but having been ſick five or ſix days, without any freſh Proviſion or other good Nouriſhment aboard, and ſeeing no likelihood of my recovery, I rather choſe to go to ſome Port in time, than to beat here any longer; my people being very negligent, when I was not upon Deck my ſelf: I found the VVinds variable, ſo that I might go any way, Eaſt, Weſt, North, or South; wherefore, its probable I might have found the ſaid Rocks, had not Sickneſs prevented me; which diſcovery (when ever made) will be of great uſe to Merchants trading to theſe parts.

From

From hence nothing material hap- *An.* 1700. pened, till we came upon the Coast of *Java.* On the 23d we saw *Princes-Isle* plain, and the Mouth of the Streights of *Sunda.* By my computation, the distance between *Timor* and *Princes-Isle,* is 14 deg. 12 min. The next day in the Afternoon, being abreast of *Creckadore* Island, I steer'd away East-North-East for an Island that lies near Mid-way between *Sumatra* and *Java,* but nearest the *Java* Shore; which is by *English* Men called *Thwart-the-way.* We had but small Winds till about three a Clock, when it freshned, and I was in good hopes to pass through before day: But at nine a Clock the Wind fell, and we got but little. I was then abreast of *Thwart-the-way,* which is a pretty high long Island; but before eleven, the Wind turned, and presently afterward it fell calm. I was then about two Leagues from the said Island; and, having a strong Current against us, before day we were driven astern four or five Leagues. In the Morning we had the Wind at North-North-West; it look'd black and the Wind unsettled: So that I could not expect to get through. I therefore stood toward the *Java* Shore, and at ten anchored in twenty four Fathom Water, black oazy Ground, three Leagues from the Shore;

An. 1700. I founded in the Night when it was calm, and had fifty-four Fathom, courſe Sand and Coral.

In the Afternoon before, we had ſeen many Proes; but none came off to us; and in the Night we ſaw many Fires aſhore. This day a large Proe came aboard of us, and lay by our ſide an hour. There were only four Men in her, all *Javians*, who ſpoke the *Malayan* Language. They ask'd if we were *Engliſh*; I anſwered, we were; and preſently one of them came aboard, and preſented me with a ſmall Hen, ſome Eggs and Coco-nuts; for which I gave ſome Beads and a ſmall Looking-Glaſs, and ſome Glaſs-Bottles. They alſo gave me ſome Sugar-canes, which I diſtributed to ſuch of my Men as were Scorbutick. They told me there were three *Engliſh* Ships at *Batavia*.

The 28th at two in the Afternoon, we anchored in twenty-ſix Fathom Water; preſently it fell calm and began to rain very violently, and ſo continued from three till nine in the Evening. At one in the Morning we weigh'd with a fine Land-wind at South-South-Eaſt; but preſently the Wind coming about at Eaſt, we anchored; for we commonly found the Current ſetting Weſt. If at any time it turn'd, it was ſo weak, that it did us

little

little good ; and I did not think it safe to venture through without a pretty brisk leading Gale ; for the passage is but narrow, and I knew not what dangers might be in the way, nor how the Tide sets in the Narrow, having not been this way these twenty-eight Years, and all my People wholly strangers : We had the Opening fair before us.

While we lay here, four *Malayan* Proes came from the Shore, laden with Coco-nuts, Plantains, Bonanoes, Fowls, Ducks, Tobacco, Sugar, *&c.* These were very welcome, and we purchased much refreshment of them. At ten a Clock I dismiss'd all the Boats, and weigh'd with the Wind at North-West. At half an hour past six in the Evening, we anchored in thirty-two Fathom Water in a course sort of Oaze. We were now past the Island *Thwart-the-way*, but had still one of the small Islands to pass. The Tide begun to run strong to the West ; which obliged me to anchor while I had Soundings, for fear of being driven back again or on some unknovvn Sand. I lay still all Night. At five a Clock the next Morning, the Tide began to slacken : At six, I vveigh'd vvith the Wind at South-East by East, a handsome Breeze. We just vveather'd the *Button* ; and sounding several times, had still be-

An. 1700. tvveen thirty and forty Fathom. When vve vvere abreaſt of the *Button*, and about tvvo Leagues from the Weſtermoſt point of *Java*, vve had thirty-four Fathom, ſmall Peppery Sand. You may either come betvveen this Iſland and *Java*, or, if the Wind is Northerly, run out betvveen the Iſland *Thwart-the-way* and this laſt ſmall Iſland.

The Wind for the moſt part being at Eaſt and Eaſt by South, I vvas obliged to run over tovvards the *Sumatra* Shore, ſounding as I went, and had from thirty-four to tvventy-three Fathom. In the Evening I ſounded pretty quick, being got near the *Sumatra* Shore; and, finding a Current ſetting to the Weſt, betvveen eight and nine a Clock vve anchored in thirty-four Fathom. The Tide ſet to the Weſt from ſeven in the Evening to ſeven this Morning; and then, having a ſmall Gale at Weſt-South-Weſt, I vveigh'd and ſtood over to the *Java* Shore.

In the Evening, having the Wind betvveen Eaſt-North-Eaſt and South-Eaſt by Eaſt, vve could not keep off the *Java* Shore. Wherefore I Anchored in twenty ſeven Fathom Water, about a League and a half off Shore. At the ſame time vve ſavv a Ship at anchor near the Shore, about tvvo miles to Leevvard of

us.

us. We found the Tide setting to the Westward, and presently after we Anchored, it fell calm. We lay still all Night, and saw many Fires ashore. At five the next Morning, being *July* the 1st, we weigh'd and stood to the North for a Sea-breeze: At ten the Wind coming out, I tack'd and had a fine brisk Gale. The Ship we saw at anchor, weigh'd also and stood after us. While we past by *Pulo Baby*, I kept sounding, and had no less than fourteen Fathom. The other Ship coming after us with all the Sail she could make, I shortned Sail on purpose that she might overtake us, but she did not. A little after five, I anchored in thirteen Fathom good oazy Ground. About seven in the Evening, the Ship that followed us, past by close under our Stern; she was a *Dutch* Fly-boat; they told us they came directly from *Holland*, and had been in their passage six Months. It was now dark, and the *Dutch* Ship anchored within a mile of us. I ordered to look out sharp in the Morning; that, so soon as the *Dutch* Man began to move, we might be ready to follow him; for I intended to make him my Pilot. In the Morning at half an hour after five we weigh'd, the *Dutch* Man being under Sail before; and we stood directly after him. At

An. 1700. eight, having but little Wind, I sent my Boat aboard of him, to see vvhat Nevvs he had brought from *Europe*. Soon after, vve spied a Ship coming from the East, plying on a Wind to speak vvith us, and shewing *English* Colours. I made a signal for my Boat, and presently bore away towards her; and being pretty nigh, the Commander and Super-cargoe came aboard, supposing we had been the *Tuscany* Galley, which was expected then at *Batavia*. This was a Country Ship, belonging to Fort *St. George*, having come out from *Batavia* the day before, and bound to *Bencola*. The Commander told me that the *Fleet-frigat* was at Anchor in *Batavia* Road, but would not stay there long: He told me also, that his Majesty Ships commanded by Captain *Warren* were still in *India*, but he had been a great while from the Coast and had not seen them. He gave me a Draught of these Streights, from the *Button* and *Cap* to *Batavia*, and shew'd me the best way in thither. At eleven a Clock, it being calm, I anchored in fourteen Fathom good oazy Ground.

At two a Clock we weigh'd again; the *Dutch* Ship being under Sail before, standing close to *Mansheters* Island; but finding he could not weather it, he tack'd and stood off a little while, and then

then tack'd again. In the mean time I ſtood pretty nigh the ſaid Iſland, ſounding, but could not weather it. Then I tack'd and ſtood off, and the *Dutch* ſtood in towards the Iſland; and weathered it. I being deſirous to have room enough, ſtood off longer, and then went about, having the *Dutch* Ship four points under my Lee. I kept after him; but as I came nearer the Iſland, I found a Tide ſetting to the Weſt, ſo that I could not weather it. Wherefore at ſix in the Evening I anchored in ſeven Fathom oazy Ground, about a mile from the Iſland: The *Dutch* Ship went about two miles further, and anchored alſo; and we both lay ſtill all Night. At five the next Morning we weigh'd again, and the *Dutch* Ship ſtood away between the Iſland *Cambuſſes* and the Main; but I could not follow, becauſe we had a Landwind. Wherefore I went without the *Cambuſſes*, and by noon we ſaw the Ships that lay at the Careening Iſland near *Batavia*. After the Land-wind was ſpent, which we had at South-Eaſt and South-South-Eaſt; the Sea-breeze came up at Eaſt. Then we went about; and the Wind coming afterward at Eaſt-North-Eaſt, we had a large Wind to run us into *Batavia* Road: And at four in the Afternoon, we anchored in ſix Fathom ſoft Oaze.

CHAP.

An. 1700.

CHAP. VI.

The A. continues in Batavia *Road, to refit, and to get Provisions.* English *Ships then in the Road. Departure from* Batavia. *Touch at the* Cape of Good Hope. *And at St.* Helena. *Arrival at the Island of* Ascension. *A Leak Sprung. Which being impossible to be stopped; the Ship is lost, but the Men saved. They find Water upon the Island. And are brought back to* England.

WE found in *Batavia* Road a great many Ships at anchor, most *Dutch*, and but one *English* Ship named the *Fleet-frigat*, commanded by one *Merry*. We rode a little without them all. Near the Shore lay a stout China Junk, and a great many small Vessels, *viz.* Brigantines, Sloops and *Malayan* Proes in abundance. Assoon as I anchored, I sent my Boat aboard the *Fleet-frigat*, with orders to make them strike their

their Pendant, which was done ſoon after the Boat went aboard. Then my Clerk, whom I ſent in the Boat, went for the Shore, as I had directed him; to ſee if the Government would anſwer my Salute: But it was now near Night, and he had only time to ſpeak with the *Ship-bander*, who told him that the Government would have anſwered my Salute with the ſame number of Guns, if I had fired as ſoon as I anchored; but that now it was too late. In the Evening my Boat came aboard, and the next Morning I my ſelf went aſhore, viſited the *Dutch* General, and deſir'd the Priviledge of buying ſuch Proviſion and Stores, as I now wanted; which he granted me.

I lay here till the 17th of *October* following, all which time we had very fair Weather, ſome Tornadoes excepted. In the mean time I ſupplied the Carpenter with ſuch ſtores as were neceſſary for refitting the Ship; which prov'd more leaky after he had caulk'd her, then ſhe was before: So that I was obliged to careen her, for which purpoſe I hired Veſſels to take in our Guns, Ballaſt, Proviſion and Stores.

The *Engliſh* Ships that arriv'd here from *England*, were firſt the *Liampo*, commanded by Captain *Monk*, bound for

An. 1700. for *China*; next, the *Panther*, commanded by Captain *Robinson*; then the *Maneel* Frigat, commanded by Captain *Clerk*. All these brought good Tidings from *England*. Most of them had been unfortunate in their Officers; especially Captain *Robinson*, who said that some of them had been conspiring to ruin him and his Voyage. There came in also several *English* Country Vessels; first a Sloop from *Ben-jarr*, commanded by one *Russel*, bound to *Bengale*; next, the *Monsoon*, belonging to *Bengale*: She had been at *Malacca* at the same time that his Majesty Ship the *Harwich* was there: Afterwards came in also another small Ship from *Bengale*.

While we stay'd here, all the forenamed *English* Ships sailed hence; the two *Bengale* Ships excepted. Many *Dutch* Ships also came in here, and departed again before us. We had several reports concerning our Men of War in *India*, and much talk concerning Rovers who had committed several Spoils upon the Coast, and in the Streights of *Malacca*. I did not hear of any Ships sent out to quash them. At my first coming in, I was told that two Ships had been sent from *Amboyna* in quest of me; which was lately confirm'd by one of the Skippers, whom I by accident met with here.

He

He told me they had three Protests against me; that they came to *Pulo-Sabuda* on the Coast of *New Guinea* twenty-eight days after my departure thence, and went as far as *Scoutens* Island, and hearing no further News of me, return'd. Something likewise to this purpose Mr. *Merry*, Commander of the *Fleet-frigat*, told me at my first arrival here; and that the General at *Batavia* had a Copy of my Commission and Instructions; but I look'd upon it as a very improbable thing.

An. 1700.

While we lay here, the *Dutch* held several Consultations about sending some Ships for *Europe* sooner than ordinary: At last the 16th of *October* was agreed upon for the day of Sailing, which is two Months sooner than usual. They lay ready two or three days before, and went out on the 10th. Their Names were, the *Ostresteen*, bound to *Zealand*; the *Vanheusen*, for *Enchiehoust*; and the three *Crowns*, for *Amsterdam*, commanded by *Skipper Jacob Uncright*, who was Commadore over all the rest. I had by this time finished my business here, *viz.* fitted the Ship, recruited my self with Provision, filled all my Water; and the time of the Year to be going for *Europe* being now at hand, I prepar'd to be gone also.

An. 1700. Accordingly on the 17th of *October*, at half an hour after six in the Morning, I weigh'd Anchor from *Batavia*, having a good Land-wind at South, and fair VVeather: And by the 19th at noon, came up with the three *Dutch* Ships before-mentioned. The 29th of *November* in the Morning we saw a small Hawk flying about the Ship till she was quite tired. Then she rested on the Mizen-Top-Sail-Yard, where we catch'd her. It is probable she was blown off from *Madagascar* by the violent Northerly Winds; that being the nighest Land to us, though distant near one hundred and fifty Leagues.

The 30th of *December*, we arrived at the *Cape of Good Hope*; and departed again on the 11th of *January* 170⁰⁄₁. About the end of the Month, we saw abundance of Weeds or Blubber swim by us, for I cannot determine which. It was all of one Shape and Colour. As they floated on the VVater, they seem'd to be of the breadth of the Palm of a Mans Hand, spread out round into many Branches about the bigness of a Mans Finger. They had in the middle a little Knob, no bigger than the top of a Mans Thumb. They were of a Smoak-colour; and the Branches, by their pliantness in the Water, seem'd to be more simple than

than Gellies, I have not ſeen the like before. An. 1700.

The 2d of *February*, we anchored in St. *Helena* Road, and ſet ſail again from thence on the 13th.

On the 21ſt we made the Iſland of *Aſcenſion*, and ſtood in towards it. The 22d between eight and nine a Clock, we ſprung a Leak, which increaſed ſo that the Chain-pump could not keep the Ship free. VVhereupon I ſet the Hand-pump to Work alſo, and by ten a Clock ſuck'd her. Then wore the Ship, and ſtood to the Southward, to try if that would eaſe her; and then the Chain-pump juſt kept her free. At five the next Morning we made Sail and ſtood in for the Bay; and at nine anchored in ten and a half Fathom, ſandy Ground. The South-point bore South-South-Weſt diſtance two miles, and the North-point of the Bay, North Eaſt half North, diſtance two miles. As ſoon as we anchored, I ordered the Gunner to clear his Powder-room, that we might there ſearch for the Leak, and endeavour to ſtop it within board if poſſible; for we could not heel the Ship ſo low, it being within four ſtreaks of the Keel; neither was there any convenient place to haul her aſhore. I ordered the Boatſwain to aſſiſt the Gunner; and by ten a Clock

the

An. 1700. the Powder-room was clear. The Carpenters Mate, Gunner, and Boatswain went down; and soon after I followed them my self, and ask'd them whether they could come at the Leak: They said they believed they might, by cutting the Cieling; I told the Carpenters Mate (who was the only person in the Ship that understood any thing of Carpenters-work,) that if he thought he could come at the Leak by cutting the Cieling without weakning the Ship, he might do it; for he had stopp'd one Leak so before; which though not so big as this, yet having seen them both, I thought he might as well do this as the other. VVherefore I left him to do his best. The Ceiling being cut, they could not come at the Leak; for it was against one of the *Foot-hook-Timbers*, which the Carpenters Mate said he must first cut, before it could be stopp'd. I went down again to see it, and found the VVater to come in very violently. I told them I never had known any such thing as cutting Timbers to stop Leaks; but if they who ought to be best judges in such cases, thought they could do any good, I bid them use their utmost Care and Diligence, promising the Carpenters Mate that I would always be a Friend to him if he could and would stop it: He said, by four a Clock in the

Afternoon

Afternoon he would make all well, it being then about eleven in the Forenoon. In the Afternoon my Men were all employ'd, pumping with both Pumps; except such as assisted the Carpenter's Mate: About one in the Afternoon I went down again, and the Carpenter's Mate was cutting the After-part of the Timber over the Leak. Some said it was best to cut the Timber away at once; I bid them hold their Tongue, and let the Carpenter's Mate alone; for he knew best, and I hop'd he would do his utmost to stop the Leak. I desir'd him to get every thing ready for stopping the violence of the Water, before he cut any further; for fear it should over-power us at once: I had already ordered the Carpenter to bring all the Oakam he had, and the Boatswain to bring all the waste Cloaths, to stuff in upon occasion; and had for the same purpose sent down my own Bed-cloaths. The Carpenter's Mate said he should want short Stantions, to be placed so that the upper-end should touch the Deck, and the under-part rest on what was laid over the Leak; and presently took a length for them. I ask'd the Master-Carpenter what he thought best to be done: He replied, till the Leak was all open, he could not tell. Then he went away to make a

An 1700 Stantion, but it was too long ; I ordered him to make many of several lengths, that we might not want of any size. So, once more desiring the Carpenter's Mate to use his utmost endeavours, I went up, leaving the Boatswain and some others there. About five a Clock the Boatswain came to me, and told me the Leak was increased, and that it was impossible to keep the Ship above VVater ; when on the contrary I expected to have had the News of the Leak's being stopt. I presently went down, and found the Timber cut away, but nothing in readiness to stop the force of the VVater from coming in. I ask'd them why they would cut the Timber, before they had got all things in readiness : The Carpenter's Mate answered, they could do nothing till the Timber was cut, that he might take the dimensions of the place ; and that there was a Chaulk which he had lined out, preparing by the Carpenter's Boy. I ordered them in the mean time to stop in Oakam, and some Pieces of Beef ; which accordingly was done, but all to little purpose : For now the Water gush'd in with such violence, notwithstanding all our Endeavours to check it, that it flew in over the Cieling ; and, for want of Passage out of the Room, over-

An. 1709.

over-flow'd it above two feet deep. I ordered the Bulk-head to be cut open, to give Paſſage to the Water that it might drain out of the Room ; and withal ordered to clear away abaft the Bulk-head, that we might bail : So now we had both Pumps going, and as many bailing as could ; and by this means the Water began to decreaſe ; which gave me ſome hope of ſaving the Ship. I ask'd the Carpenter's Mate, what he thought of it ; He ſaid, *Fear not ; for by ten a Clock at Night I'll engage to ſtop the Leak.* I went from him with a heavy Heart ; but putting a good Countenance upon the Matter, encouraged my Men, who pump'd and bail'd very briskly ; and, when I ſaw occaſion, I gave them ſome Drams to comfort them. About eleven a Clock at Night, the Boatſwain came to me, and told me, that the Leak ſtill encreaſed ; and that the Plank was ſo rotten, it broke away like Dirt ; and that now it was impoſſible to ſave the Ship ; for they could not come at the Leak, becauſe the Water in the Room was got above it. The reſt of the Night we ſpent in Pumping and Bailing. I worked my ſelf to encourage my Men, who were very diligent ; but the Water ſtill encreas'd, and we now thought of nothing but ſaving our Lives. Wherefore I hoiſted out the

An. 1700. Boat; that, if the Ship should sink, yet we might be saved: And in the Morning we weighed our Anchor, and warp'd in nearer the Shore; yet did but little good.

In the Afternoon, with the help of a Sea-breeze, I ran into seven fathom, and anchored; then carried a small Anchor ashore, and warp'd in till I came into three fathom and a half. Where having fastned her, I made a Raft to carry the Mens Chests and Bedding ashore; and, before eight at Night, most of them were ashore. In the Morning I ordered the Sails to be unbent, to make Tents; and then my self and Officers went ashore. I had sent ashore a Puncheon, and a 36 Gallon Cask of Water, with one Bag of Rice for our common use: But great part of it was stolen away, before I came ashore; and many of my Books and Papers lost.

On the twenty-sixth following, we, to our great comfort, found a Spring of fresh Water, about eight miles from our Tents, beyond a very high Mountain, which we must pass over: So that now we were, by God's Providence, in a condition of subsisting some time; having plenty of very good Turtle by our Tents, and Water for the fetching. The next day I went up to to see the Watering-place, accompanied with most of my Officers. We lay by the way all Night, and next Morning early got thither; where we found a very fine spring on the South-East side of the high Mountain, about half a mile from its top: But the continual Fogs make it so cold here, that it is very unwholsome living by the Water. Near this place, are abundance of Goats and Land-crabs. About two mile South-East from the Spring, we found three or four shrubby Trees, upon one of which was cut an Anchor and Cable, and the Year one thousand six hundred and forty-two. About half a Furlong from these, we found a convenient place for sheltering Men in any Weather. Hither many of our Men resorted; the hollow Rocks affording convenient Lodging; the Goats, Land-crabs, *Men of War Birds*, and Boobies, good Food; and the Air was here exceeding wholsome.

About a Week after our coming ashore, our Men that liv'd at this new Habitation, saw two Ships making towards the Island. Before Night they brought me the News; and I ordered them to turn about a score of Turtle, to be in readiness for their Ships if they should touch here: But before Morning they were out of sight, and

An. 1700. the Turtle were releas'd again. Here we continued without seeing any other Ship till the second of *April*; when we saw eleven Sail to Windward of the Island: But they likewise past by. The Day after appear'd four Sail, which came to anchor in this Bay. They were his Majesty's Ships the *Anglesey*, *Hastings* and *Lizard*; and the *Canterbury East-India* Ship. I went on board the *Anglesey* with about thirty-five of my Men; and the rest were dispos'd of into the other two Men of War.

We sail'd from *Ascension*, the 8th; and continued aboard till the 8th of *May*: At which time the Men of War having miss'd St *Jago*, where they design'd to Water, bore away for *Barbadoes*: But I being desirous to get to *England* as soon as possible, took my passage in the Ship *Canterbury*, accompanied with my Master, Purser, Gunner, and three of my superiour Officers.

THE

Fishes taken on the Coast of **New Guinea.**

This Fish his fins & Taill is Blew. wth Blew spots all over ye Body.

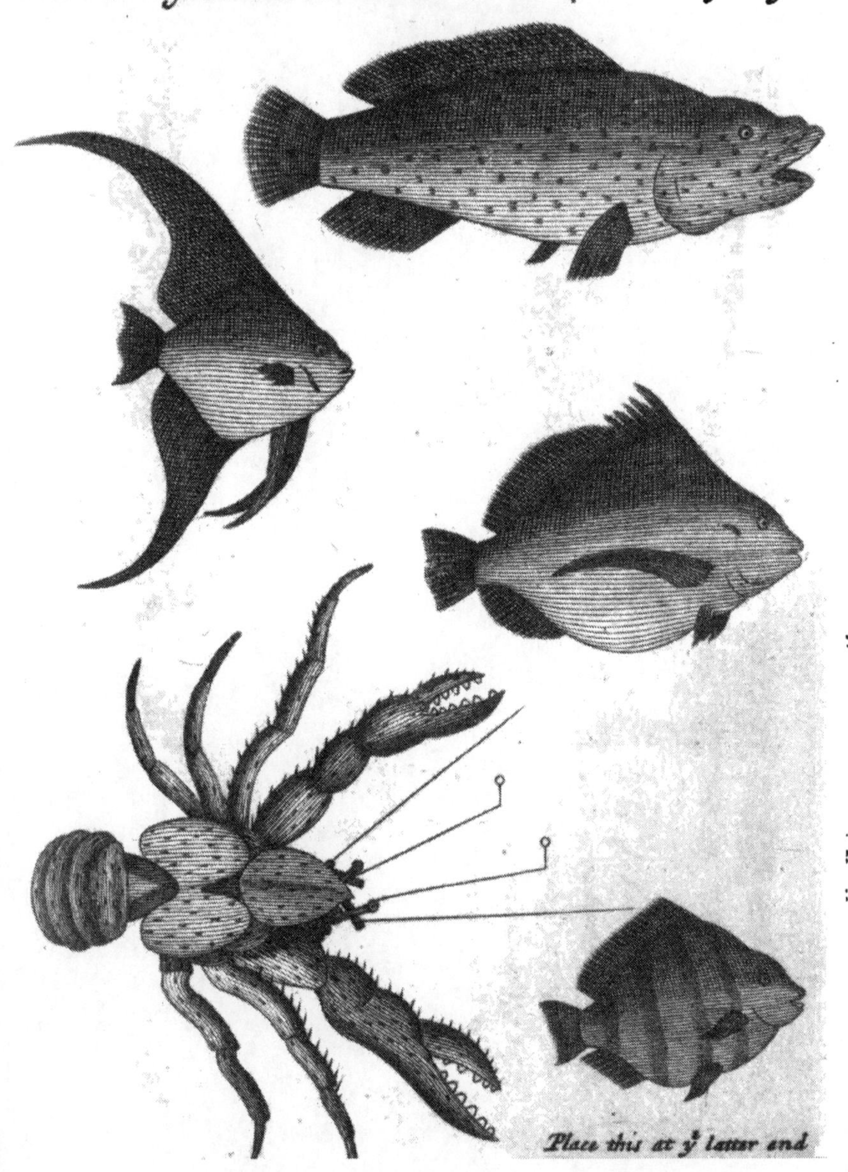

The Mountain Cow;
or, as some think:
The Hippopotamus:
Described in Capt. Dampier
2d Vol. in Campeachy in
Page 102. 3. 4. 5. 6 & 7.

THE

INDEX.

A.

B.

O 4 *Bird,*

C.

D.

E.

F.

G

J.

K.

L.

M.

N.

O.

P.

Pidgeons,

T.

Printed in the USA
CPSIA information can be obtained
at www.ICGtesting.com
LVHW011048281024
794948LV00001B/31